The Emerging Business Models

Singapore University of Social Sciences - World Scientific Future Economy Series

ISSN: 2661-3905

Singapore University of Social Sciences - World Scientific Future Economy Series introduces the new technology trends and challenges that businesses today face, financial management in the digital economy, blockchain technology, smart contract and cryptography. The authors describe current issues that the business leaders and finance professionals are facing, as well as developments in digitalisation. The series covers several increasingly important new areas such as the fourth industrial revolution, Internet of Things (IoT), blockchain technology, artificial intelligence (AI) and many other forces of disruption and breakthroughs that shape today's realities of the economy. A better understanding of the changing environment in the future economy can enable business professionals and leaders to recognise realities, embrace changes, and create new opportunities — locally and globally — in this inevitable digital age.

*Published**

Vol. 3 *Artificial Intelligence, Data and Blockchain in a Digital Economy, First Edition*
edited by David Lee Kuo Chuen, supported by Singapore University of Social Sciences and World Scientific, in support of Singapore Digital (SG:D) and in collaboration with Infocomm Media Development Authority

Vol. 2 *The Emerging Business Models*
by Guan Chong, Jiang Zhiying and Ding Ding

Vol. 1 *AI & Quantum Computing for Finance & Insurance: Fortunes and Challenges for China and America*
by Paul Schulte and David Lee Kuo Chuen

Forthcoming Titles

Financial Management in the Digital Economy
Ding Ding, Guan Chong and David Lee Kuo Chuen

*More information on this series can also be found at
https://www.worldscientific.com/series/susswsfes

(Continued at end of book)

Singapore University of Social Sciences - World Scientific Future Economy Series 2

The Emerging Business Models

GUAN Chong
JIANG Zhiying
DING Ding
Singapore University of Social Sciences, Singapore

SUSS SINGAPORE UNIVERSITY OF SOCIAL SCIENCES

World Scientific

Published by

World Scientific Publishing Co. Pte. Ltd.
5 Toh Tuck Link, Singapore 596224
USA office: 27 Warren Street, Suite 401-402, Hackensack, NJ 07601
UK office: 57 Shelton Street, Covent Garden, London WC2H 9HE

Library of Congress Cataloging-in-Publication Data
Names: Guan, Chong, author. | Jiang, Zhiying (College teacher), author. |
Ding, Ding, 1980– author.
Title: The emerging business models / Guan Chong, Jiang Zhiying, Ding Ding.
Description: Singapore ; Hackensack, NJ : World Scientific, [2020] |
Series: Singapore university of social sciences - world scientific future economy series, 2661-3905 ; 2 | Includes bibliographical references and index.
Identifiers: LCCN 2019031027 | ISBN 9789811203923 (hardcover) |
ISBN 9789811204401 (paperback) | ISBN 9789811203930 (ebook)
Subjects: LCSH: Electronic commerce. | Electronic commerce--Social conditions. |
Disruptive technologies.
Classification: LCC HF5548.32 .G849 2020 | DDC 658.8/72--dc23
LC record available at https://lccn.loc.gov/2019031027

British Library Cataloguing-in-Publication Data
A catalogue record for this book is available from the British Library.

For any available supplementary material, please visit
https://www.worldscientific.com/worldscibooks/10.1142/11372#t=suppl

Desk Editor: Yulin Jiang

Typeset by Stallion Press
Email: enquiries@stallionpress.com

About the Authors

Dr David LEE Kuo Chuen is a Professor at the Singapore University of Social Sciences (SUSS) and Vice President of the Economic Society of Singapore. He is the founder of several entities including BlockAsset Ventures, the external advisor to Stanford University's Distributed Trust Initiative, a Fulbright Scholar, and an Independent Director of listed and technology companies. He graduated with BSc, MSc, and Ph.D. from the London School of Economics and Political Science in Econometrics. He was the Director of the Sim Kee Boon Institute for Financial Economics, Group Managing Director of OUE and Auric Pacific. He founded the Ferrell hedge fund in 1998 and was a property developer. He has more than 60 cited articles and nine books on Asian economics, digital currency, blockchain, and inclusive FinTech. His latest book is entitled *AI & Quantum Computing for Finance & Insurance: Fortunes and Challenges for China and America*.

Dr GUAN Chong is an Associate Professor and Head, Marketing Programme in the School of Business, Singapore University of Social Sciences (SUSS), Singapore, where she has been a faculty member since 2011. Dr Guan completed her Ph.D. in Marketing at Nanyang Business School, Nanyang Technological University, and her undergraduate studies in Marketing at Guanghua School of Management, Peking University, China. Her research interests lie in the area of consumer decision-making on digital platforms. Her publications appear in leading journals, such as *Journal of Business*

Research, *Telecommunications Policy*, and *Psychology & Marketing*. She has taught various marketing courses at undergraduate and graduate levels.

Dr JIANG Zhiying is a senior lecturer from the Marketing Department. Before joining the Singapore University of Social Sciences (SUSS), she worked as an Assistant Professor in the School of Economics, Erasmus University Rotterdam. Her areas of expertise include Marketing Analytics, Marketing Research, Mobile Economics, and Product Design. She tends to use quantitative methods to identify the underlying trends and opportunities under various micro or macroenvironments. In addition, Dr Jiang has consulting experience with multinational firms such as Philips and IMS Health during her years in the Netherlands. She was also an expert appointed by the European Union Commission under H20 framework on issues, such as optimal marketing expenditure in the pharmaceutical industry.

Associate Professor DING Ding is the Vice Dean in the School of Business at the Singapore University of Social Sciences (SUSS). She graduated from Nanyang Technological University with a Ph.D. in economics and is also a Chartered Financial Analyst (CFA). She has rich experience in teaching economics and finance courses to undergraduate, post-graduate, and MBA students, and she has also conducted executive training programs for SME owners and senior management. Recently, she has developed and taught the first course in Financial Technologies and Innovations at SUSS to undergraduate students. Her research interests lie in the areas of financial technology and digital inclusion, economy of China, and international capital markets.

Dr WANG Yue is the Head, Doctor of Business Administration Programme in the School of Business, Singapore University of Social Sciences (SUSS). Before she joined SUSS, she graduated from the National University of Singapore with a Ph.D. in Marketing from the School of Business and Master of Research in Economics from the Economics Department, Faculty of Arts and Social Sciences. She was trained with various quantitative approaches. Her research interests center around digital marketing, social media, and new product adoption. She teaches digital marketing-related courses in both undergraduate and post-graduate levels, as well as executive training programs.

LEE Zhan Xiong Sean is an adjunct lecturer in the School of Business, Singapore University of Social Sciences (SUSS) and a consultant at Kantar. Mr Lee graduated from the Master of Science Management Programme at the National University of Singapore (NUS). His research interests lie in the area of digital marketing and social media. His publications appear in the *World Review of Business Research* and several international conference proceedings. At Kantar, Mr Lee specializes in trends and futures research, helping clients develop future-proof strategies and frameworks. He has handled key client accounts, such as Google, P&G, Unilever, Kraft Heinz, Danone, and Beam Suntory.

Dr LAU Jing Feng is the Head, Mathematics Programme in the School of Science and Technology, Singapore University of Social Sciences (SUSS), Singapore, where he has been a faculty member since 2016. He graduated from Purdue University with a Ph.D. in mathematics and his undergraduate studies in the National University of Singapore (NUS). Prior to joining SUSS, he has held research fellow appointments in Tel Aviv University and NUS. He has extensive experience in teaching mathematics courses to undergraduate and post-graduate students. Recently, he has developed interests in machine learning and data science.

Contents

Chapter 1

The Fourth Industrial Revolution (Industry 4.0)

1.1 Introduction

Technological revolution has been changing the way we live, work, and connect with one another, at the scale, scope, and complexity that is unlike anything humankind has experienced before. The First Industrial Revolution leveraged on water and steam power to mechanize production. The Second harnessed electric power for mass production. The Third used electronics and information technology to automate production. Building on the Third, the Fourth Industrial Revolution is characterized by a fusion of advanced technologies that is disrupting almost every industry in every economy.

The breadth and depth of these technological breakthroughs herald the transformation of entire ecosystems of stakeholders, from the public and private sectors to academia and governance. The Fourth Industrial Revolution builds upon the digital innovations, representing a wide range of new technologies that are transforming both the digital and the physical worlds, and impacting all industries and economies. Professor Klaus Schwab, founder and executive chairman of the World Economic Forum, describes in his book, *The Fourth Industrial Revolution*, how this major industrial era has fundamental differences from the previous three industrial revolutions.

1.2 Stages of Industrial Revolutions

1.2.1 *The First Industrial Revolution*

The First Industrial Revolution took place from about 1760 to 1840 in Europe and America. This transition included introduction of mechanization and manufacturing with water and steam power. It was a period when many agrarian societies became industrial and urban economies. The textile and iron industries, along with the development of the steam engine and the invention of other machine tools (e.g., screw cutting lathe, cylinder boring machine and the milling machine), played important roles in the industrial revolution.

1.2.2 *The Second Industrial Revolution*

The Second Industrial Revolution took place between late 19th to early 20th centuries. During this period, introduction of division of labor and mass production via electric power enabled growth for pre-existing industries and expansion of new ones, such as steel, telephone, light bulb, and electricity. This revolution gradually expended to the chemical industries, petroleum refining and distribution sector, and, in the 20th century, the automotive industry. The increasing supply of economical petroleum products also reduced the importance of coal and further realized the potential for industrialization. By the 1890s, the first giant industrial corporations like US Steel, General Electric, Standard Oil and Bayer AG were created.

1.2.3 *The Third Industrial Revolution*

The Third Industrial Revolution began in the 1960s. It is usually called the computer or digital revolution because it was catalyzed by the development of semiconductors, mainframe computing (1960s), personal computing (1970s and 80s) and the Internet (1990s). It refers to the advancement of technology from analogue electronic and mechanical devices to the information technologies (ITs). Advancements during the Third Industrial Revolution include the operation of electronics and IT for the automation of production, personal computers and the internet.

1.2.4 *The Fourth Industrial Revolution*

Today, we are in the midst of a fourth industrial revolution that began at the start of the 21st century and builds on the digital revolution. The Fourth Industrial Revolution builds upon the Third Industrial Revolution, representing emerging technology breakthroughs in a number of fields, including analytics, robotics, artificial intelligence, virtual and augmented reality, nanotechnology, high performance computing, biotechnology, the Internet of Things (IoT), 3D printing and autonomous vehicles. These advanced technologies have great potential to improve the efficiency of business and organizations.

There are three reasons why today's transformations represent not merely a prolongation of the Third Industrial Revolution but rather the arrival of a Fourth and distinct one: velocity, scope, and systems impact. The speed of current breakthroughs has no historical precedent. When compared with previous industrial revolutions, the Fourth is evolving at an exponential rather than a linear pace. Moreover, it is disrupting almost every industry in every country. And the breadth and depth of these changes herald the transformation of entire systems of production, management, and governance (Schwab, 2015).

Table 1.1 shows the areas of focus in the Fourth Industrial Revolution. Increasing trends in artificial intelligence point to significant economic disruptions in the coming years. Artificial systems that rationally solve complex problems pose a threat to many kinds of

Table 1.1: Fourth Industrial Revolution Areas of Focus

Areas of Focus
• AI and Machine Learning • IoT and Connected Devices • Blockchain and Distributed Ledger Technology (DLT) • Autonomous and Urban Mobility • Drones and Tomorrow's Airspace • Precision Medicine • Digital Trade and Cross-Border Data Flows

Source: World Economic Forum (2018).

employment, but also offer new avenues to economic growth. A report by McKinsey & Company found that half of all existing work activities would be automated by currently existing technologies, thereby enabling companies to save billions of dollars and to create new types of jobs. For example, driverless cars may modestly replace tax and Uber drivers, but autonomous trucks may radically transform shipping with far fewer jobs for truck drivers.

In addition, innovative technologies will integrate different scientific and technical disciplines. Key forces will come together in "a fusion of technologies that is blurring the lines between physical, digital, and biological spheres" (Schwab, 2015). This fusion of technologies goes beyond mere combination. Fusion is more than complementary technology, because it creates new markets and new growth opportunities for each participant in the innovation. It blends incremental improvements from several (often previously separated) fields to create a product.

As the physical and digital worlds continue to connect and converge, new technologies and platforms will increasingly enable citizens to engage with governments and voice their opinions. Simultaneously, governments can harness new technological tools to improve their interaction.

For instance, a Smart City is an urban development vision that integrates various information and communication technologies (ICTs), artificial intelligence (AI), robotics, IoT, and data analytics solutions into the planning and management of a city, to improve quality of life, efficiency of urban operation and competitiveness. Table 1.2 shows the key parameters of a smart city.

Underpinning the growing smart cities are a host of broad ICT trends:

- The development of advance networking and communication technologies (enables connection among infrastructure, devices, and individuals; gather data and deliver services to targeted endpoints);
- Digital–physical systems (the digital representation of physical devices to the Internet);

Table 1.2: Smart City Key Parameters

Governance and Education	• eGovernment • eEducation
Smart Energy	Digital Management of Energy • Smart grids • Smart meters • Intelligent energy storage
Smart Healthcare	• Use of eHealth and eHealth systems • Intelligent and connected medical devices • Smart lifestyle choices
Smart Security	• Surveillance • Biometrics • Simulation modeling and crime protection • Disaster management solutions
Smart Infrastructure	• Sensor networks • Digital water and waste management
Smart Building	Automated intelligent buildings • Building automation • Advanced lighting automation
Smart Mobility	Advanced traffic management system (ATMS) • Parking management • IT-enabled transportation pricing system • Use of green mobility options

Source: Frost & Sullivan. Retrieved from http://www.egr.msu.edu/~aesc310web/resources/SmartCities/Smart%20City%20Market%20Report%202.pdf.

- Cloud computing (the delivery of computing as a service, has offered organizations ways to reduce costs and increase efficiency);
- Open data (encourages public agencies to release data sets and make them available for public usage).

1.3 Challenges and Opportunities

Like the revolutions that preceded it, the Fourth Industrial Revolution has the potential to raise global income levels and improve the quality of life for populations around the world. To date, those who have gained the most from it are the consumers who are able to afford and access the digital world; technology has brought about new products and services that increase the efficiency and pleasure of our personal lives. Ordering a cab, booking a flight, buying a product, making a payment, listening to music, watching a film, or playing a game — any of these can now be done remotely.

In the future, technological innovation will also lead to a supply-side miracle, with long-term gains in efficiency and productivity. Transportation and communication costs will drop, logistics and global supply chains will become more effective, and the cost of trade will diminish, all of which will open new markets and drive economic growth.

1.3.1 *The Impact on Business*

The Fourth Industrial Revolution is having a major impact on businesses. The speed of innovation and the velocity of technological disruption are hard to anticipate and predict. In many industries, the introduction of new technologies is creating entirely new ways of serving existing customer needs. These innovations have significantly disrupted existing industry value chains.

From the business perspective, many companies are seeing more and more agile and innovative competitors who leverage on global digital platforms for research and development, marketing, and distribution and improve their product quality, service speed, or price at which value is delivered.

From the consumer perspective, with growing information transparency and digital consumer engagement, consumers are now demanding more customized touch-points, virtual or physical, forcing businesses to adapt the way they design, market, and deliver goods and services. For example, the "sharing" or "on demand" economy combines both demand and supply to disrupt existing industry structures.

A key trend is the development of technology-enabled platforms that combine both demand and supply to disrupt existing industry structures, such as those we see within the "sharing" or "on demand" economy. These technology platforms, rendered easy to use by the smartphone, convene people, assets, and data — thus creating

Figure 1.1: Key Impacts That the Fourth Industrial Revolution Has on Businesses

entirely new ways of consuming goods and services in the process. In addition, they lower the barriers for businesses and individuals to create wealth, altering the personal and professional environments of workers. These new platform businesses are rapidly multiplying into many new services, ranging from laundry to shopping, from chores to parking, from massages to travel. Figure 1.1 summarizes the key impacts that the Fourth Industrial Revolution has on businesses.

Overall, the inexorable shift from simple digitization (the Third Industrial Revolution) to innovation based on combinations of technologies (the Fourth Industrial Revolution) is forcing companies to re-examine the way they do business. The bottom line, however, is the same: business leaders and senior executives need to understand their changing environment, challenge the assumptions of their operating teams, and relentlessly and continuously innovate.

1.3.2 *The Impact on Individuals*

The Fourth Industrial Revolution will change not only what we do but also who we are. It will affect our identity and all the issues associated with it: our sense of privacy, our notions of ownership, our consumption patterns, the time we devote to work and leisure, and how we develop our careers, cultivate our skills, meet people, and nurture relationships. It is already changing our health and leading to a "quantified" self, and it may lead to human augmentation in the near future.

For example, robotics can and will change our lives in many aspects. Technically robots are automated motorized tools. They cook food, play our music, record our shows, and even run our cars. Consequently, robots have the potential to improve the quality of our lives at home, work, and many other places. Customized robots will create new jobs, improve the quality of existing jobs, and give people more time to focus on what they want to do.

1.3.3 *The Impact on Government*

As the physical, digital, and biological worlds continue to converge, new technologies and platforms will increasingly enable citizens to engage with governments, voice their opinions, coordinate their efforts, and even circumvent the supervision of public authorities. Simultaneously, governments will gain new technological powers to increase their control over populations, based on pervasive surveillance systems and the ability to control digital infrastructure. Overall, governments will increasingly face pressure to change their current approach to public engagement and policymaking, as their central role of conducting policy diminishes owing to new sources of competition and the redistribution and decentralization of power that new technologies make possible.

Ultimately, the ability of government systems and public authorities to adapt will determine their survival. If they prove capable of embracing a world of disruptive change, subjecting their structures to the levels of transparency and efficiency that will enable them to maintain their competitive edge, they will endure. If they cannot evolve, they will face increasing trouble.

This will be particularly true in the realm of regulation. Current systems of public policy and decision-making evolved alongside the Second Industrial Revolution, when decision-makers had time to study a specific issue and develop the necessary response or appropriate regulatory framework. The whole process was designed to be linear and mechanistic, following a strict "top down" approach.

In the future, such an approach is no longer feasible. Given the Fourth Industrial Revolution's rapid pace of change and broad impacts, legislators and regulators are being challenged to an unprecedented degree. It is important to understand how they could preserve the interest of the consumers and the public at large while continuing to support innovation and technological development. The answer is to embrace "agile" governance, just as the private sector has increasingly adopted agile responses to software development and business operations more generally. This means regulators must continuously adapt to a new, fast-changing environment, re-inventing themselves so they can truly understand what it is they are regulating. To do so, governments and regulatory agencies will need to collaborate closely with business and civil society.

As this process takes place and new technologies such as autonomous or biological weapons become easier to use, individuals and small groups will increasingly join states in being capable of causing mass harm. This new vulnerability will lead to new fears. At the same time, advances in technology will create the potential to reduce the scale or impact of violence, through the development of new modes of protection, for example, or greater precision in targeting.

1.4 Challenges of the Fourth Industrial Revolution

The evolution of global industries in the Fourth Industrial Revolution is both exciting and challenging. Life will change with 3D printing, IoT, and the fusion of technologies. The Fourth Industrial Revolution can raise income levels by allowing entrepreneurs to "run" with their new ideas. It will improve the quality of life for many people around the world. Consumers are likely to gain the most from the Fourth Industrial Revolution.

While there are many benefits of the Fourth Industrial Revolution, there are several key challenges that lie ahead. At the same time, the revolution could yield greater inequality, particularly in its potential to disrupt labor markets. As automation substitutes for labor across the entire economy, the net displacement of workers by machines might exacerbate the gap between returns to capital and returns to labor. The scarcest and most valuable resource in an era driven by digital technologies will be neither ordinary labor nor ordinary capital; rather it will be those people who can create new ideas and innovations. In the future, talent, more than capital, will represent the critical factor of production. People with ideas, not workers or investors, will be the scarcest resource. The quest for talent will give rise to a job market that may become increasingly segregated. Low skilled and low wage jobs will be replaced by computers and digitization. The higher paid jobs requiring more skills are less likely to be replaced. This increased dichotomization can lead to an increase in social tensions.

In addition to the threat of massive job displacement under the ongoing Fourth Industrial Revolution, there are a variety of challenges, such as cybersecurity, hacking, risk assessment, and others. A higher level of alert is raised when our lives become extensively connected to various devices, from our cell phones, cars, and light switches to our home security cameras, and smart speakers. One of the biggest trends we are currently experiencing is that everything is connected and there is no going back.

Having everything attached to everything else in the IoT is going to monumentally increase the vulnerabilities present in any given network. With more knobs, connections, and burden of connectivity, systems have to be more secure. The Fourth Industrial Revolution calls for greater cybersecurity. Companies will need to map their networks, assessing the risk and critical factors relating to security. Such an assessment should examine accessibility to systems, such as possible threats from internal sources, from disgruntled employees to internal human error, and external sources including hackers and cyber terrorists. Further, companies must assess risk and determine if these risks will be accepted, reduced, shared via insurance or other vehicles, or rejected.

Risks can be from both intentional and unintentional sources. If your house lights turn on via your computer, but you have lost the wireless connection to your house, you may be living in the dark. Unintentional sources of risk can include errors promulgated by company employees or nature itself, such as storms causing disruptions in connectivity. Individuals too should assess their risks, just as companies will. It may happen that the Internet will have more information about individuals than the family, friends, and colleagues of the individuals. Certainly, the ability of data to be processed and the speed in which it can be done surpass the ability and speed of individuals. It is necessary to examine the value of processes and assets, from machinery to intellectual property, ensuring that vulnerabilities identified had been rectified.

When we consider the changing nature of security threats — from employees connecting personal devices to company networks to brute force attacks from hackers — the situation is further complicated. The sophistication in risk identification and neutralization has to change with it. While data can be lost or stolen by employees, either inadvertently or intentionally, the biggest attacks in recent years have been external malicious attacks, collectively or commonly referred to as hacking. These could be hacking to move money around, such as when Russian hackers stole US$10 million from Citibank customer accounts, Internet terrorism, such as the US$2 million damages caused to WeaKnees.com over a six-week botnet attack, Internet pump-and-dump fraud where hackers take advantage of manipulating stock prices, or software piracy which is estimated to cost over US$50 billion a year.

The Fourth Industrial Revolution is more than just technology-driven change. Rather, it is powered with disruptive innovation to positively impact our core industries and sectors, such as education, health, and business.

In education, with the previous industrial revolutions, the focus of education has changed. With the First Industrial Revolution, education was focused on standard modes of learning, such as the McGuffey reader, with move toward mass production in the Second Industrial Revolution and standardized testing. Education is service-oriented and with the move into the Third Industrial Revolution, we come to see students under a customized learning model. Now in the Fourth Industrial Revolution, technologies really blur the lines between physical, digital, and biological spheres. Disruptive innovation makes its way into higher education in which it redefines the conventional ways universities deliver their content to students. New modes of curriculum and teaching arise, and the focus changes from modes of teaching to modes of learning. Alternative curriculums are being constantly developed. Disruptive innovation also re-shapes how businesses operate. Thinking has really moved outside of the box.

Finally yet importantly, in an era featuring AI, automation, robots, and genetic engineering, we have new ethical concerns emerging. Many debates have arisen in genetic engineering about the use of tools and research technologies. On one hand, preventing genetic disease by genetic engineering is desirable. On the other hand, what guidelines, or regulation, or ethical boundaries we should establish in order to prevent the over manipulation genetics for desirable traits? Is there such a thing as over manipulation? Infused with AI and machine learning ability, robots have become smarter and more autonomous, but they still lack an essential feature — the capacity of moral reasoning. This limits their ability to make good or ethical decisions in complex situations. Further, the most critical question is whose moral standards robots should inherit. Moral values differ greatly from individual to individual, across countries, religions, and ideological boundaries. Uncertainty over which moral framework to adopt underlies the difficulty and limitations to ascribing moral values to artificial systems.

1.5 Summary

As industrial revolutions have moved from the mechanization of production in the First Industrial Revolution, to the mass production in the second, and then to the automation of production in third, the standards of living for most people around the world have greatly improved. Undoubtedly, the capability of advancing technology coming from the latest industrial revolution has the potential to make even bigger and greater improvements on every aspect of our lives — changes than the first three industrial revolutions summed together.

On the other hand, there are a variety of challenges stemming from the Fourth Industrial Revolution to overcome. From income inequality to cybersecurity, the benefits of the Fourth Industrial Revolution have obstacles that must be harnessed, directed and overcome, such as income inequality, cybersecurity, and ethical dilemmas. Technology and advancements in science drive transformation around the world. They create ripple effects on societies, institutions, and economies. They will transform the ways in which we live, work, and interact with one another. Understanding these new technologies and their disruption potential is critical for all nations and especially developing countries.

The Fourth Industrial Revolution may affect society and economy in a variety of ways. First, a large portion of people around the world are likely to use social media platforms to connect, learn, and change information. Second, a variety of innovative producers and competitors will have easy access to digital platforms of marketing, sales, and distribution, thereby improving the quality and price of goods and services. Third, consumers will be more and more involved in the production and distribution chains. The main effects of this revolution on the business environment are the impact it will have on consumer expectations, product quality, the move toward collaborative innovation, and innovations in organizational forms.

References

Bloem, J., Doorn, M. V., Duivestein, S., Excoffier, D., Maas, R., & Ommeren, E. V. (2014). The Fourth Industrial Revolution. *Things Tighten.*

The Boston Consulting Group (BCG) (2017). *Is UK Industry Ready for the Fourth Industrial Revolution?* Retrieved from https://media-publications.bcg.com/Is-UK-Industry-Ready-for-the-Fourth-Industrial-Revolution.pdf.

Deloitte (2018). *The Fourth Industrial Revolution Is Here — Are You Ready?* Deloitte Insights. Retrieved from https://www2.deloitte.com/content/dam/insights/us/articles/4364_Industry4-0_Are-you-ready/4364_Industry4-0_Are-you-ready_Report.pdf.

Schwab, K. (2015). *The Fourth Industrial Revolution: What It Means and How to Respond.* Retrieved from https://www.foreignaffairs.com/articles/2015-12-12/fourth-industrial-revolution.

World Economic Forum (2018). Centre for the Fourth Industrial Revolution. Retrieved from https://www.weforum.org/centre-for-the-fourth-industrial-revolution/areas-of-focus.

Chapter 2

Internet of Things

2.1 Introduction

In recent years, the Internet of Things (IoT) has received tremendous attention from individuals and businesses all around the world due to its potential to fundamentally shift the way we interact with our surroundings. IoT is the network of wireless communication devices, including physical devices, home appliances, vehicles, and other computer items embedded with electronics, software, sensors, actuators, and connectivity which enables these items to connect and exchange data. "Things," in the IoT context, can refer to a mixture of hardware, software, data and service, ranging from devices such as mobile phones, wearable devices, automobiles with built-in sensors, traffic cameras, etc. and their embedded computing system. These devices collect useful data, exchange the data among other devices and produce real-time analytics and enable machine learning. The convergence of multiple "things" across existing network infrastructure creates opportunities for more direct integration of the physical world into Internet-based systems, resulting in improved efficiency, accuracy and economic benefits and reduced human intervention. The relationship can be among people–people, people–things, and things–things. One of the strongest benefits of IoT is the efficiency it can offer (see Table 2.1).

Table 2.1: Size of the IoT Market Worldwide in 2014 and 2020 by Industry (in billion US dollars)

Market in Billion US Dollars	2014	2020
Consumer electronics	1150	2225
Automotive	850	1780
Healthcare and life sciences	520	1335
Industrial/manufacturing	472	890
Construction/infrastructure	—	890
Retail	—	890
Energy and utilities	—	890

Source: Columbus (2017).

2.2 IoT Applications

The most discussed applications for IoT include smart cities, environment, water, metering, security and emergency services, retail, logistics, industrial control, agriculture, farming, domestic and home automation, and e-health. In a research paper published by McKinsey Global Institute (MGI) in 2015, they have identified nine settings, capturing IoT use in environments such as homes, offices, factories, worksites (mining, oil and gas, and construction), retail environments, cities, vehicles, and the outdoors. Among all the settings, there is also a "human" setting for systems that attach to the human body and enable such health and wellness applications to monitor chronic disease or exercise, and productivity-enhancing applications such as use of augmented reality technology to guide workers in performing complex physical tasks (see Table 2.2).

2.2.1 *Human*

Two types of IoT technology applications fall under the human setting. The first category is health and fitness. The second set — human productivity — involves using IoT technology to improve performance in the workplace. IoT has the potential for transformative change in human health. Using connected devices to continuously monitor patients as they live their lives — particularly those with chronic conditions such as diabetes — the IOT can improve

Table 2.2: Settings of IoT Applications

Setting	Description	Examples
Human	Devices attached to or inside the human body	Devices (wearables and ingestibles) to monitor and maintain human health and wellness; disease management, increased fitness, higher productivity
Home	Buildings where people live	Home controllers and security systems
Retail environments	Spaces where consumers engage in commerce	Stores, banks, restaurants, arenas — anywhere consumers consider and buy; self-checkout, in-store offers, inventory optimization
Offices	Spaces where knowledge workers work	Energy management and security in office buildings; improved productivity, including for mobile employees
Factories	Standardized production environments	Places with repetitive work routines, including hospitals and farms; operating efficiencies, optimizing equipment use and inventory
Worksites	Custom production environments	Mining, oil and gas, construction; operating efficiencies, predictive maintenance, health and safety
Vehicles	Systems inside moving vehicles	Vehicles including cars, trucks, ships, aircraft, and trains; condition-based maintenance, usage-based design, pre-sales analytics
Cities	Urban environments	Public spaces and infrastructure in urban settings; adaptive traffic control, smart meters, environmental monitoring, resource management
Outside	Between urban environments (and outside other settings)	Outside uses include railroad tracks, autonomous vehicles (outside urban locations), and flight navigation; real-time routing, connected navigation, shipment tracking

Source: *The Internet of Things: Mapping the Value beyond the Hype.* McKinsey Global Institute, June 2015.

patient adherence to prescribed therapies, avoid hospitalizations (and post-hospitalization complications), and improve the quality of life for hundreds of millions of patients. This could have an economic impact of US$170 billion to US$1.6 trillion per year in 2025. The use of IoT systems could enable societal benefits worth more than US$500 billion per year, based on the improved health of users and reduced cost of care for patients with chronic diseases.

Human productivity applications include the use of augmented reality devices, such as goggles through which data can be displayed to guide the performance of factory workers. The goggles would present information such as instructions for physical tasks, which would appear to float in the worker's field of vision, allowing the worker to refer to the correct procedures without having to find a computer terminal. Using IoT data, companies can also redesign jobs and processes for greater efficiency and effectiveness. In addition, IoT technology can help mobile workers in the field to stay connected and work more effectively. Together these applications could have an impact of US$150–350 billion globally in 2025 (including the size of the human productivity benefits in the settings in which they can be achieved).

2.2.2 *Home*

A wide range of IoT devices and applications are emerging for use in the home, including connected thermostats, smart appliances, and self-guided vacuum cleaners.

As these devices evolve, we expect that the greatest economic impact from the IOT in the home will be in chore automation, which we estimate can cut 100 hours of labor per year for the typical household. That could be worth nearly US$135 billion globally in 2025. The next largest impact would come from energy management (up to US$110 billion per year), followed by security, which would have an impact of more than US$20 billion per year, based on injuries and deaths avoided. In total, it is estimated that IoT applications in the home could have an economic impact of US$200–350 billion per year in 2025.

2.2.3 *Offices*

Offices are defined as the physical environments in which knowledge work is the primary activity. Key benefits of IoT use in office settings are in security and energy management. By using digital security cameras with advanced image-processing capabilities, operators of office buildings can monitor activity throughout their properties without requiring guards to patrol or continuously monitor video feeds. It is estimated that IoT-based energy management in offices could cut energy use by 20%. In total, office IoT applications could have an economic impact of US$70–150 billion per year in 2025.

2.2.4 *Factories*

This setting is one of the largest sources of value from the adoption of IoT, potentially generating an economic impact of US$1.2–3.7 trillion per year. Factories can be defined in the broadest sense to include all standardized production environments. Therefore, the estimated benefits of IoT use include the use in hospitals and in agricultural settings, as well as in manufacturing facilities. In the factories setting, value from IoT would arise mainly from productivity improvements, including 10–20% energy savings and a 10–25% potential improvement in labor efficiency. Improvements in equipment maintenance, inventory optimization, and worker health and safety are also sources of value in factories.

2.2.5 *Worksites*

Worksites are defined as custom production environments, such as mines, oil and gas extraction sites, and construction sites. Leading companies that operate in worksite settings were early adopters of IoT technology. A typical oil-drilling platform today might use 30,000 sensors, watching over the performance of dozens of systems.

In mining, self-driving vehicles, including mine cars and ore trucks, are helping to streamline operations and reduce costs. Overall, improvements in operations from IoT applications could be worth more than US$470 billion per year in 2025 in worksites.

The second major source of value — potentially more than US$360 billion per year — would be improved equipment maintenance. Using sensors to monitor the health of machinery in use, companies can shift to a condition-based maintenance model (maintaining equipment when there is an actual need through predictive analytics) rather than relying on a regular maintenance schedule or repairing equipment only when it breaks down.

At the same time, companies can also improve human health and safety by using IoT. In total, it is estimated that IoT in the worksites setting can have an economic impact of US$160–930 billion per year in 2025.

2.2.6 *Vehicles*

In the vehicles setting, we can look at the potential for IoT to monitor and improve the performance of planes, trains, and other vehicles while in use, which could generate US$210–740 billion per year in IoT impact for this setting in 2025.

2.2.7 *Cities*

Cities have become the focus of a great deal of innovation and experimentation with IoT technology, through the so-called smart city initiatives. Since cities are the engines of global economic growth — the 600 largest cities in the world are expected to generate 65% of global gross domestic product (GDP) growth through 2025 — the impact of IoT technologies can be substantial. Specifically, cities can benefit from IoT in four areas: transportation, public safety and health, resource management, and service delivery.

Transportation is the largest application and includes IoT-based systems to manage traffic flow and autonomous vehicles (self-driving cars). For example, there is great economic potential in the use of IoT that could come from adjusting commuting schedules based on actual tracking data of public transit systems (buses and trains). Up to 70% of commuting time today is "buffer time" — the extra time between when the rider arrives at a stop or station and when the bus or train actually leaves. Reducing the buffer in cities across the world could provide time savings equivalent to more than US$60 billion per year. In total, IoT transportation applications could be worth more than US$800 billion per year to cities around the world.

The next biggest impact would be in public health — up to nearly US$700 billion per year, mainly from air and water quality improvements that would reduce lives lost to pollution. Using IoT smart meters to reduce loss of electricity in distribution and sensors to detect water leaks could be worth as much as US$69 billion per year globally. Overall, it is estimated that IoT application in the cities setting could have an economic impact of US$930 billion to US$1.6 trillion per year in 2025.

2.2.8 *Outside*

The outside setting captures uses of IoT technology outside all of the other settings, that is, those that take place outdoors between urban environments. For example, it includes the use of IoT to improve the routing of ships, airplanes, and other vehicles between cities using advanced navigation informed by various sensors. This also includes using IoT to track containers and packages in transit. It is estimated that these applications could have an economic impact of US$560–850 billion per year in 2025 (see Table 2.3).

Table 2.3: Top IoT Industries and Subsectors

Hardware	Software	Integrated Solutions	Consulting
• Semiconductors • Connectivity • Measurement sensors • Power batteries	• Data analytics software • Firmware devices/sensors • Platform	• Monitoring and controlling • Positioning and tracking • Production optimization • Risk management • Security	• Consulting services

Source: Frost & Sullivan (2017).

2.3 What Makes IoT Possible?

The emergence of IoT depends on a variety of technologies as essential building blocks and their simultaneous operation and communication among one another. Objects/devices are equipped with sensors and actuators and given the capability to emit, accept, and process signals. The tagging of things may be achieved through technologies like Radio-frequency identification (RFID), near field communication, barcodes, QR codes and digital watermarking.

The essential building blocks of an IoT architecture include the following:

- **Application, application programming interfaces (APIs) and user interaction**
  - Collaboration involving users, applications and business processes.
- **Cloud server**
  - Computing systems/platforms such as enterprise cloud servers with application hosting, real-time processing, big data storage and management capacity, predictive analytics, artificial intelligence (AI), machine learning, and content delivery.
- **Network connectivity**
  - Ultra-high speed Internet access through wireless/wired methods, including Wi-Fi, Bluetooth, Virtual Private Network (VPN), Cellular 2G/3G/4G.
- **Gateway**
  - Communication standards and protocols that enable connectivity between the sensors, the network, and the physical objects/devices.
- **Sensors**
  - Sensors measure physical inputs and convert the information from the physical environment into a signal, which is then digitally storable for access and analysis. Sensors can measure anything from temperature, force, pressure, traffic flow and position, to light intensity.

- Sensors are being embedded into everything, from electricity networks, roads and other infrastructure, to mobile, wearable, home automation and security devices.
- When physical objects/devices are equipped with identifiers, computers can store/manage their signals and transmit information via the Internet.

- **Actuators**
  - Act on the signal from the sensors and convert it into output.

2.4 Unlocking the Potential of the Internet of Things

With the wide range of connected devices using embedded software/sensors to communicate, collect and exchange data with one another, IoT offers a virtually endless array of opportunities. Human–machine interactions can unlock massive potential of IoT to achieve:

2.4.1 *Real-time Measures of Business Performance and Data-driven Business Decisions*

With billions of connected devices transmitting valuable real-time market information, companies can measure performance of each business event and function within the organization accurately. These new sources of data streams and predictive capabilities are driving companies to re-evaluate their current business decisions.

2.4.2 *Operational Efficiency and Workforce Productivity*

By using IoT to automate business and manufacturing processes, and remotely monitor and control operations, companies can optimize supply chains, conserve resources and improve productivity.

Wearables and other IoT-enabled devices can help employees improve decision-making, automate routine tasks, fast-track communication, and more. For example, companies can deliver real-time insights on how their fleet is being operated. Connecting fleets and cargo to IoT empowers companies to optimize their recourses on the road with an improved estimated time of arrival (ETA) forecasting. IoT enables businesses to track driver behaviors and utilize data analytics to perform predictive maintenance. By looking into data on their logistics processes, companies can make informed decisions and leverage predictive intelligence that adds value and optimizes their supply chains. Thus, these technologies can boost workforce productivity and job satisfaction.

2.4.3 *New Business Models and Revenue Stream*

IoT is disrupting traditional business models and creating massive new opportunities for companies to create new products and services based on real-time sensor data and information, such as automatic product replenishment.

In addition to gaining operating efficiencies and the other benefits that companies can derive from using IoT in their existing businesses, companies can create new business models based on IoT:

- **New pricing models:** IoT creates a continuous, fine-grained stream of real-time data. These data enable both microsegmentation and insights into consumer willingness to pay. Customized, dynamic pricing models based on specific customer preferences, as well as the usage context, can provide benefits for both companies and consumers. For example, auto insurance companies have begun to sell policies with premiums based on sensor data about actual driving behavior, rather than the demographics of drivers. A company selling smart lighting fixtures could charge customers based on the amount of money they save.
- **Service-based business models:** The Internet has enabled the "as a service" business model for IT infrastructure and software. IoT enables "anything as a service" business models for all kinds of other products, potentially letting many kinds of companies shift from selling products to selling services based on those products.

This model can transform large capital expenditures into a pay-by-usage operating expense. Examples of this trend are proliferating. They include selling "power by the hour" rather than gas turbines, selling transportation services rather than automobiles, and selling pages printed rather than laser printers. IoT technology not only enables the providers of these services to charge by usage, but it also enables these companies to better maintain and upgrade the equipment that is used in these services, removing the maintenance burden from customers and creating the basis of long-term relationships. IoT-based service businesses can also use consolidated customer usage data and data about individual customers for cross-selling, guiding product development, and other purposes. Furthermore, IoT can allow products to become better while in service (the opposite of depreciating in value), a concept Hal Varian has dubbed "product kaizen".

2.4.4 *Monetization of IoT Data*

The exhaust data generated by IoT applications can become a profit center itself. For example, almost any data describing consumer behavior can be of great value to marketers. Data about physical assets such as buildings and vehicles can be used to assess insurance risks. One company's data exhaust could be another company's gold mine, and that value could be monetized by the originator of the data. Of course, privacy, confidentiality, and ownership rights over such data — for instance, consumer purchasing data — are issues that would have to be addressed to create such business models.

2.4.5 *Enhanced Customer Experiences*

IoT breaks down the barrier between the customer and the organization and drives innovations that create exceptional consumer experiences. The increasing sophistication of the sensors embedded in technology makes it possible for devices ("things") to read, gauge, and understand consumers at unprecedented levels.

From integrating customer care with actual product performance and usage, to delivering highly personalized products and services, IoT offers many ways to create more compelling customer experiences across digital and physical worlds. Digital identity tracking, seamless authentication of rights and content access, increased content personalization, contextual customer profiling and automated recommendations are important for the execution of IoT.

For example, personal mobile devices that can link with their daily appliances have already been introduced to consumers. Alphanumeric passwords have long existed as a means to control access and security to technology. However, they are not a biologically intuitive way to extract and authenticate identity. A less intrusive and more secure method would come in the form of sensors that can read personal attributes such as a fingerprint, face, or a heart rate. This is the type of human–machine interaction that becomes easier with IoT and can solve for use cases including over-the-top (OTT) authentication.

2.5 Barriers and Challenges

Although IoT applications have great potential, certain conditions must exist to enable the full economic benefits of IoT in the above-mentioned settings: machinery that includes IoT sensors and actuators; short-range and long-range data communications networks with sufficient reliability and capacity; and improvements in data analytics. Data security and confidentiality are important to address, as are organizational issues concerning workforce and collaboration.

Fortunately, after four decades of factory automation, introducing IoT into the factory environment in many cases will not require wholesale replacement of production equipment since many machines already in use have sensors or can be retrofitted with the needed electronics. This is in sharp contrast to earlier industrial waves, such as the introduction of steam, which required replacing more than 80% of installed production machinery.

Still, because factories are capital-intensive and the rate of equipment replacement can be very low (especially in developing economies), the upgradability of current machinery for IoT applications is a critical enabler of adoption. So, while most equipment could be upgraded to become IoT-ready, adoption could still be gradual, especially for small- and medium-sized manufacturing enterprises. We estimate that the adoption rate of IoT in manufacturing in 2025 can reach 65–90% in advanced economies and 50–70% in developing economies. These levels of penetration are similar to those seen for factory automation equipment over the past few decades.

Another requirement for adopting IoT broadly is the reliable data networks with sufficient capacity. Data networks within factory settings must often operate in environments with large amounts of electromagnetic interference.

The continuous flow of data between machines and remote computer systems in IoT setups also requires high-bandwidth, long-distance communications. In many instances, particularly in developing economies, plants are in remote locations, perhaps several hundred kilometers from major cities that have advanced telecom infrastructure.

Maximizing the benefits of IoT-based systems also depends on improvements in the analytics — algorithms that can interpret and act on the flow of real-time data from many machines. In today's environment, little of the data generated by production machinery is actually used for decision-making. Better analytics would help companies use more of the information they collect for optimization and prediction.

Data security and confidentiality issues also affect IoT adoption. To gather the data for usage-based design improvements, manufacturers need access to data about how their customers are using products. This may raise questions about confidentiality since a manufacturer is likely to consider specific details about factory performance to be confidential. This concern needs to be overcome if IoT impact is to be maximized.

The final barriers to adoption of IoT in factory environments involve human resources and organizational issues. First is the lack of capabilities in most factory organizations to utilize IoT systems and data. A skilled workforce is needed to take full advantage of IoT technologies. Furthermore, if IoT data remain siloed in manufacturing organizations or if different departments do not have incentives to collaborate, the maximum benefits will not be realized. For example, companies cannot improve the quality of their products if performance data collected by the service department are not shared freely with the product development group. Obstacles to data sharing need to be overcome if IoT impact is to be maximized.

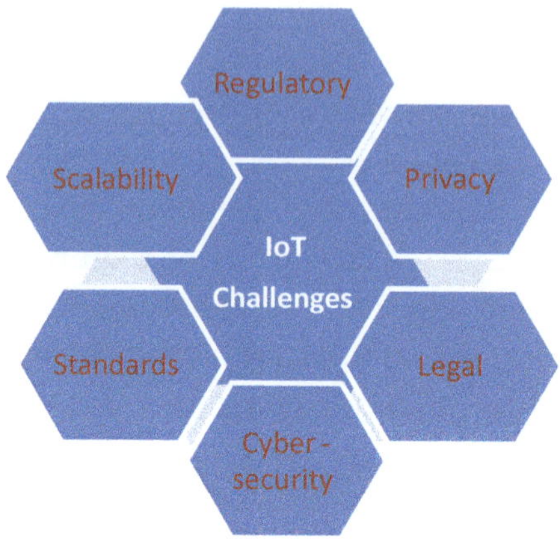

2.5.1 *Regulatory Hurdles*

Around the world, legislators are trying to plug the regulatory holes that IoT is creating. In late 2014, the European Commission published an opinion on IoT specifically focused on wearables and smart home devices. Among its recommendations was the requirement that users remain in complete control of their data; the opinion also addressed the activities that organizations need to consider to remain compliant with European Union data protection laws. Early in 2015 in the US, Congresswoman Suzan DelBene (D-WA) and Congressman Darrell Issa (R-CA) announced the launch of a Congressional caucus on IoT. Its purpose is to educate members about both the opportunities and challenges IoT poses for health, transportation, home and work, and subsequently to find the balance between collecting data that IoT creates and protecting consumer privacy. Around the same time, the Federal Trade Commission released a report that recommended data minimization and self-regulatory programs to improve privacy and security practices.

2.5.2 *Privacy in the Age of IoT*

Companies are going to be faced with massive amounts of data that all of these devices are going to produce. Companies need to figure out a way to store, track, manage, analyze and make sense of the vast amounts of data that will be generated.

In almost every consumer survey, privacy emerges as the No.1 concern. In an IDC US consumer survey, close to 55% of respondents identified "ensuring my privacy" as their top expectation of third-party providers of home automation services. Similarly, a Forrester's survey of global enterprise decision-makers identified security and privacy among the top five concerns to IoT adoption and growth. Privacy is a major challenge that organizations need to overcome, as the IoT ecosystem seeks to collect enormous amounts of data and contextual inputs from sensors and other IoT solutions.

2.5.3 *New Targets for Hackers Increase Pressure on Cybersecurity*

With billions of devices being connected simultaneously, IoT also opens up companies all over the world to more security threats. Prioritizing security during the IoT design process, bolstering public privacy protections, advancing transparency and openness are important steps to make sure that the entire network stays secure.

Cybersecurity remains a leading challenge for consumers and businesses alike — something that will become exponentially more difficult as IoT connects more devices, software, machines and humans. In a recent EY publication *Creating Trust in the Digital World: Global Information Security Survey 2015*, 88% of study respondents indicate that they do not believe their information security fully meets the organization's needs. However, when asked about IoT, 68% do not consider monitoring their business ecosystem as an information security challenge. Similarly, 67% do not see managing the growth in access points to their organization as a challenge to their information security. These numbers would seem to indicate that many organizations underestimate the impact IoT may have on their business.

2.5.4 *The Setting of New Legal Precedents*

IoT lives in the physical world. Operational failures resulting from connectivity loss or device malfunction are inevitable. These failures also create legal risks for companies. Lawyers and judges across jurisdictions are grappling with some tough legal questions:

- Who is responsible for a connected device malfunction or resulting accident?
- Who is responsible for a data breach?
- How much are companies liable vs. the consumers themselves?

As individual cases make their way through the legal system, many of these questions remain unanswered.

2.5.5 *Intellectual Property Rights*

One of the biggest questions on everyone's mind is: Who owns the data? Is it the company that manufactures the sensor, or the company that manufactures the device? Or the individual whose data are being measured and collected?

Certainly, European Union (EU) legislators are emphasizing the rights of the individual to own their data, but this is not the case in all jurisdictions. Even in instances where data ownership is clear, the duration for which owners can own the rights of collected data still needs to be addressed.

2.5.6 *IoT Standards*

For IoT to progress, connectivity standards need to evolve. Similar to the communication challenges that people from different parts of the world face, in today's IoT ecosystem, devices, sensors, machines and people are often speaking completely different languages with one another. Without a common language or standard of implementation, IoT will remain limited in its application.

2.5.7 *Scaling to Reach Critical Mass*

In the same way, the IoT ecosystem needs standard protocols and a single language to expand. It also needs consistency in device architecture for large-scale adoption. Only when IoT reaches critical mass can companies reap the full financial benefits of their personalized content investments.

2.6 Implications for Consumers

IoT offers substantial benefits for consumers, as well as a new set of risks. On the benefit side, IoT applications offer the potential to drive down the costs and enhance the quality of goods and services, as manufacturers, retailers, and other suppliers employ IoT systems to improve product design and operations. Additionally, as mentioned in an earlier analysis, another source of value for consumers will be greater consumer convenience and time savings. As they travel, consumers can benefit from IoT-managed roadways, self-driving cars, real-time public transit information, and planes that take off and land on schedule. In stores, shoppers could experience a new level of service, savings via targeted in-store offers, and automated checkout. At home, they can offload housework to smart appliances, save money on energy use, and protect themselves with IoT-enabled security systems. And most importantly, lives can be extended and improved by IoT health and wellness applications.

Given the value associated with interoperability, consumers can choose to purchase systems that are interoperable so that they can maximize benefits. In some cases, this might mean purchasing different systems from one vendor; in others, it might mean purchasing systems from multiple vendors that use common standards.

IoT benefits also come with a new range of risks. The Internet has already raised concerns about consumer privacy, and the mobile Internet (smartphones) has heightened those concerns by collecting passive tracking data. IoT devices will go beyond that. Wellness monitors can not only report where a person is but also gather data about what the person is doing. Health IoT systems will collect information about the consumer's weight, eating habits, exercise regimes, drugs used, and diseases. In the workplace, IoT systems can be used to monitor and manage employee performance, including by "reading" facial expressions to assess a worker's emotional state.

Consumers will need to be diligent about monitoring the data that are being gathered about them. When they sign up for services, they should bear in mind what kind of permissions they are granting for the use of their data, as well as what kind of access they allow to the data they are generating. Consumers can also participate in policymaking dialogues as regulators determine how detailed personal information can be used and how people can exert control over their personal data.

Finally, consumers will have to manage information overload. The physical and psychological effects of constant Internet and smartphone use are beginning to be acknowledged, as are "context switching costs" when the brain is interrupted by an email alert or an incoming tweet. Consumer IoT systems can be yet another source of overwhelming data and distraction. Consumers will need to think about which IoT systems to adopt and how to integrate the use of those systems into their lives. Many may avoid systems that provide a constant stream of information and instead opt for those that present information only when it is relevant.

2.7 Implications for Companies

Companies building strategies around IoT will need to develop new organizational capabilities and, in many cases, adopt different behaviors. For companies to make the most of IoT technology, their organizations must embrace data-driven decision-making capabilities and create environments in which relevant information is made available to the right decision-makers (which could be people or algorithms) in a timely manner. This may require training workers in new skills and investing in new software systems to capture and analyze IoT data. Indeed, all of the implications of big data and analytics apply to IoT data.

Whatever the application, the value of IoT data depends on effective use. Organizations will need to become adept at data-driven decision-making, which requires analytical rigor. As companies build up their data analytics capabilities (both technologies and talent) to handle the flood of IoT data, the analytics experts must be connected with decision-makers as well as frontline managers; both groups need to learn how to apply insights from large, real-time data streams in a timely manner. And when real-time action needs to be taken on a large scale — such as optimizing machine control across an entire factory or providing real-time marketing offers to shoppers — organizations will have to develop processes that enable algorithms to make decisions automatically, while allowing decision-makers to monitor metrics and set policy.

IoT will also transform the role of the IT function within organizations. The scope of responsibilities of IT departments will no longer be primarily in data centers and managing desktop and mobile devices. IoT applications will extend IT systems into physical assets, facilities, and inventory. Thus, more than ever, the IT organization will have to be closely aligned and even integrated with the operating parts of the business. And the metrics of success for IT will have to measure more than technology, extending to the performance of the business itself.

Synthesizing data from different sources will take more than interoperable systems (not a given); it will also require collaboration among leaders and teams in different parts of the organization. Owners of various IT systems (chief financial officers and chief marketing officers, for example) must be receptive to linking up their systems and providing data access across the organization.

Many IoT applications will also require collaboration. For example, in health care, IoT data will need to be shared among doctors, hospitals, insurance companies, device manufacturers, and software providers. In some instances, IoT applications may even depend on competitors working together. In mining, for example, multiple small operators would need to share data in order to optimize their operations in the same way as a large operator. To get insights from data on the characteristics of an ore body in one mine and make effective predictions requires a large comparative dataset, which small operators may not be able to get unless they pool their data.

References

Columbus, L. (2017). *Forbes 2017 Roundup of Internet of Things Forecasts.* Retrieved from https://www.forbes.com/sites/louiscolumbus/2017/12/10/2017-roundup-of-internet-of-things-forecasts/#31b923521480

Frost & Sullivan (2019). *Smart Cities, Frost & Sullivan Value Proposition.* Retrieved from https://ww2.frost.com/wp-content/uploads/2019/01/SmartCities.pdf.

McKinsey Global Institute (2015). *The Internet of Things: Mapping the Value beyond the Hype. McKinsey, June 2015.* Retrieved from https://www.mckinsey.com/~/media/McKinsey/Business%20Functions/McKinsey%20Digital/Our%20Insights/The%20Internet%20of%20Things%20The%20value%20of%20digitizing%20the%20physical%20world/The-Internet-of-things-Mapping-the-value-beyond-the-hype.ashx

Chapter 3

Artificial Intelligence and Cloud Computing

Lee Zhan Xiong Sean and Lau Jing Feng

3.1 Introduction

With technology playing a big part in everyone's life in today's world, something that was not even deemed possible 30 years ago, one can only wonder what the future holds for businesses and individuals alike in the development of the current available technologies. Two big developments that have the potential to disrupt the market and change the way we work is Artificial Intelligence and Cloud Computing. In this chapter, we will take a look at the definition, application and future impact of both of these technologies that are playing a bigger role in global companies today.

3.2 Artificial Intelligence and Its Applications in Industry

Artificial Intelligence (AI) is the simulation of human intelligence processes by machines. It has gained tremendous progress due to the wide applications to various industries, such as finance, healthcare, retail and transportation.

Industry in general views AI positively and believes that adopting AI will provide business opportunities and growth. The recent industrial AI boom has been mainly due to the fact that sensors required for the automated data acquisition have become more affordable and the advent of better hardware has greatly boosted the computation capabilities of computers to perform more complex tasks at faster speed with lower cost. In many business settings, AI is used to do routine and low-level processes, such as scanning and analyzing documents and spotting patterns in consumer preferences to boost sales.

The purpose of this expository chapter is to give an introduction to AI and highlight some aspects of how AI is being used in various industries to improve existing products or services to consumers, process applications for greater productivity improvement, and also provide insight to business models.

Artificial Intelligence is an amalgamation of three prolific technologies: machine learning, natural language processing, and cognitive computing.

Machine learning is the science of getting computers to learn, act like humans do and improve their learning over time in an autonomous fashion by feeding them data in the form of observations and real-world interactions. This learning is achieved by statistical methods.

Natural Language Processing (NLP) is a field of artificial intelligence that enables computers to analyze and understand human language. It was formulated to build software that generates and comprehends natural languages, so that a user can have natural conversations with his or her computer instead of communicating through programming languages.

Cognitive computing is the study of simulation of human thought processes and problem solving in a computerized model. Using self-learning algorithms in data mining and pattern recognition, computers can mimic the way humans think.

We now proceed to give an overview of the more commonly used algorithms used in machine learning. Under machine learning, there are two types of learning: supervised learning and unsupervised learning.

Supervised machine learning algorithms are machine learning algorithms which require external assistance. Each such algorithm learns a function that maps an input to an output variable based on examples of input and output. The input dataset consists of the train dataset and test dataset. Algorithms of this kind try to pick up some kind of patterns from the training dataset and apply them to the test dataset, either for prediction or classification.

For supervised learning, three commonly adopted techniques are Naïve Bayes, Support Vector Machine (SVM) and k-nearest Neighbors. Naïve Bayes is usually used for classification and it falls back on conditional probability. Trees are created based on the probability of events. SVM is usually used for classification and works by drawing margins such that the distance between the nearest data points and the margin is maximized so that classification error is minimized (Figure 3.1).

k-nearest neighbors algorithm assigns an object the class most common among its k-nearest neighbors or a value to an object, which is the average of the values of its k-nearest neighbors.

Unsupervised machine learning algorithms only have input data and no corresponding output variables and the objective of such algorithms is to model the underlying structure or distribution in the input data so as to learn more about the data. Two common examples

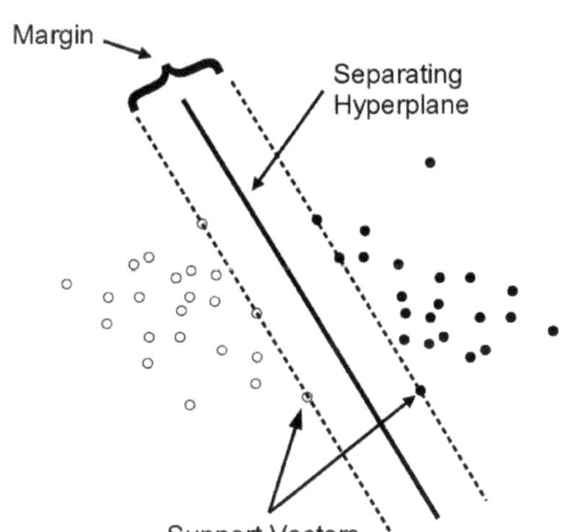

Figure 3.1: Linear Separable Case Classification

Source: Meyer (2012)

of such algorithms include the k-means clustering and Principal Component Analysis (PCA). k-means clustering is used when the training data is unlabelled and the objective is to find k groups in the data where data points are clustered based on feature similarity. PCA is a method which changes a set of data with variables into a set of linear variables not correlated using an orthogonal transformation whereby the dimension of the dataset is reduced to facilitate faster computations.

3.2.1 *AI Applications in Finance Industry*

Financial institutions such as banks and insurance companies have started to adopt AI in various operations.

Chatbots and virtual personal assistants are used to interact with customers and to solve problems before engaging human staff. American Express have used chatbots for customer service as well as payments. Customers of Kasisto are even able to use chatbots to check and monitor their spending habits.

Routine jobs such as processing loans or insurance underwriting is easily handled by machine learning. This is especially true for large commercial businesses such as big banks or insurance companies which can leverage on huge amounts of consumer data. The underlying trends and risks can be evaluated by machine learning and actuarial algorithms to determine whether the borrowers will default or how likely the insurer will actually need to pay compensation. Results from such analysis reap great returns for such companies which can afford to recruit data scientists.

AI is also extensively used for portfolio management, to automate and execute trades for investment. It was not till the last five years that robo-advisors have sprung up. They provide a means for investors to passively invest and manage their investment portfolio by keying in their profile, financial goals, as well as the amount they invest by charging a certain fraction of fees. The automated system then suitably adjusts to the respective investor's goals, risk appetite and market conditions. In Singapore, besides the robo-advisors provided by several banks, there are also other ones regulated by the Monetary Authority of Singapore (MAS) such as StashAway, AutoWealth and Smartly.

AI methods have been extensively used for trading but with great varying degrees of success. AI algorithms allow computers to use available data to predict future trends and outcomes. However, there are serious drawbacks in trying to forecast accurately the prices of securities this way due to several reasons:

- When financial markets do not exhibit any stable patterns, financial models based on AI algorithms fail to predict correctly.
- AI algorithms do not take into account unforeseen circumstances that can cause prices of securities to change unpredictably.

Using AI techniques, computing systems are able to investigate anomalies leading to fraud and flag these out to security teams. Six fraud and security AI executives from TechEmergence that were interviewed all believed that "genuinely" learning systems will be necessary in the next five to 10 years to prevent security breaches.

3.2.2 *AI Applications in Healthcare Industry*

AI has started to provide many applications to healthcare and it is widely believed that these applications will be further developed and improved.

Using artificial intelligence, AI robots can analyze data from pre-op medical records as well as past operations and leverage on this to assist surgeons to perform surgery. This approach reduces errors and complications as compared to surgeons operating without the aid of such robots. Mazor Robotics have created AI-assisted robots to assist in orthopedic surgery which has been shown in a study on 379 orthopedic patients to reduce surgical complications by five-fold. Further analysis conducted in the context of orthopedic surgery have exhibited that AI-assisted robotics when applied properly to orthopedic surgery can reduce a patients' length of stay in the hospital after surgery by 21%, which results in annual savings of US$40 billion.

Healthcare industries have started to implement AI for patient diagnosis and predict future mutations in patients. There have been promising results so far. A study conducted in Stanford University pitted an AI algorithm against 21 dermatologists to recognize skin cancers and clinical findings proves that the AI algorithm works as well as the dermatologists in identifying skin cancers. Baidu Research has even announced that its deep learning algorithm for breast cancer detection can outperform human pathologists.

Without appealing to AI technology, comparing 3D medical scans currently can take two hours or more. A research team led by Massachusetts Institute of Technology (MIT) came up with a machine learning algorithm which can do this 1,000 times faster and enables surgeons to study changes in malign conditions almost in real-time. It is expected that artificial intelligence can make future radiology equipment accurate and good enough to produce radiological images, which can replace the need for physical tissue samples obtained through biopsies.

Similar to the finance industry, AI is also employed to automate routine administrative tasks which then allow healthcare providers to be more efficient in their work and focus on other important tasks. The company Nuance offers AI solutions to assist healthcare providers to reduce documentation time and improve reporting quality by providing real-time clinical documentation guidance so that providers can ensure they receive accurate clinical history and consistent recommendations.

3.2.3 *AI Applications in Retail Industry*

AI is applied by retail companies to utilize the data of shopping experience of customers to understand shopping preferences of customers and improve their shopping experiences, both online and in traditional brick and mortar shops. By incorporating AI into their advertising, companies will be able to suitably adjust their marketing strategy. This approach has enabled companies to increase their sales.

On 23 January 2018, Amazon opened a cashier-less, queue-less convenience store Amazon Go in downtown Seattle. Customers can walk directly into the convenience store, take any item they want and walk out without paying a cashier. They pay by logging into a specific phone app. Using sophisticated machine vision and other AI applications, Amazon can track every object in the store in real-time. The opening of such stores led people to identify several issues: Such "walk-in" stores sometimes attracted lengthy queues formed outside the respective stores and some customers had not been charged for taking some items. It remains to be seen whether we can integrate AI with other technology to resolve these problems to come up with stores without queues and checkouts.

3.2.4 *AI Applications in Logistics, Supply Chain and Transportation Industry*

Autonomous vehicles such as autonomous buses, cars, drones, trains, trucks and even ships are one of the most promising applications of AI.

Google, Uber, Tesla and other companies have all ventured into the business of manufacturing and selling autonomous cars. However, to the best of the author's knowledge, level 5 autonomous driving which is complete automation has not been achieved so far.

Currently, the technology will need a lot more refinement before autonomous vehicles are deemed to be safe and reliable. However, if accident rates of such driverless vehicles can be reduced to levels much lower than any human driver, we will probably see mass adoption of these modes of transportation.

Previously, goods transported by trucks, rail and ships were tracked by satellite using telematics and optimization has been used to minimize costs and transit times. This has improved greatly because the additional computing power provided by various IoT and artificial intelligence platforms enables leading supply chain and transportation companies to mine the huge amount of data they have for real-time decision-making and predictive analysis.

Certainly, AI technology also poses challenges in its widespread adoption. This can be largely attributed to two main points: the increased security risk due to the dependence of data which may be sensitive, as well as AI causing certain jobs which were traditionally done by humans to become redundant.

3.3 Cloud Computing: Our World in the Clouds

> "I don't need a hard disk in my computer if I can get to the server faster... carrying around these non-connected computers is byzantine by comparison."
>
> —Steve Jobs

Cloud computing has taken the world by storm recently, as more and more companies, both big and small, migrate their technology ecosystems onto the cloud. In fact, the market is so huge that Gartner predicts that Worldwide Cloud Computing services may reach a value of US$411 billion by 2020 (Gartner, 2017). Amazon itself, number 18 on the Fortune 500 list in 2018, counts on its Amazon Web Services to drive 10% of their total revenue and 89% of their total operating income (Richter, 2019). The numbers are even more striking when looking at what cloud computing does for businesses and the economy. IDC (2017)'s research has shown that Salesforce, one of the biggest cloud computing services company in the world, will be creating 3.3 million new jobs by 2022 and US$859 billion worth of revenue for businesses that engage in cloud computing. To put that in context, for every US$1 spent on cloud computing services with Salesforce, the ecosystem it has in place for all its clientele will make US$5.18. It is not all about numbers, as the amount of wastage of old servers and abandoned hardware with an increase in the number of jobs has a tremendous global impact and will benefit everyone engaging in cloud computing services.

3.3.1 *What Is Cloud Computing?*

With the impact of cloud computing becoming clear, the question next is, what exactly is cloud computing? Google defines it as "the practice of using a network of remote servers hosted on the Internet to store, manage, and process data, rather than a local server or a personal computer", while Amazon, a big player in the cloud computing space, defines it as "the on-demand delivery of compute power, database storage, applications, and other IT resources through a cloud services platform via the internet with pay-as-you-go pricing". For the purpose of this chapter, we will adopt the official definition from the National Institute of Standards and Technology (NIST), which defines cloud computing as:

> "A model for enabling ubiquitous, convenient, on-demand network access to a shared pool of configurable computing resources (e.g., networks, servers, storage, applications, and services) that can be rapidly provisioned and released with minimal management effort or service provider interaction."

3.3.2 *Origins of Cloud Computing*

Cloud computing can be traced back to the 1960s, where IBM and DEC had Remote-Job Entry, the first of its kind where a program had the capability of time-sharing. About 10 years later, full-time sharing capabilities across platforms was available on programs such as Multics. Telecommunication companies in the 1990s started the first form of cloud computing, with their offerings of Virtual Private Network (VPN) services. These services enabled the Telcos to dedicate band-width appropriately to heavy users and light users, which preceded the concept of "cloud".

In the 2000s, cloud computing became its own and truly found worldwide fame. Amazon released its Amazon Web Services in 2006, while Google followed closely in 2008 with the Google App Engine. The creation of these services resulted in Gartner publishing a report that predicted the rise and need of cloud computing in the future "to shape the relationship among consumers of IT services, those who use IT services and those who sell them" (Gartner, 2010). The same report noted that organizations began reducing the number of owned hardware and software and switching to a pay-per-use model online. With this, the birth of cloud computing services became a heavily adopted phenomenon.

3.3.3 *Characteristics*

With cloud computing still a new concept, NIST has identified five key characteristics that define cloud computing:

1. **On-demand self-service:** Consumers can access computing capabilities as needed without the requirement of a human to provision these services from the provider company.

2. **Broad network access:** The capabilities of the service is accessed through a network through heterogeneous client platforms (such as mobile phones, PCs, laptops, tablets).
3. **Resource pooling:** Resources such as network bandwidth, storage and memory, and processor power are pooled by the provider to serve multiple consumers, adjusting the allocation by assigning and re-assigning according to the consumer demand for the service at any one point of time.
4. **Rapid elasticity:** Capabilities need to be able to be provisioned and released according to usage scale, so as to provide a sense of unlimited capability to consumers whenever they need it at any time.
5. **Measured service:** Cloud systems need to be able to automatically optimize and control the resources available to their consumers by leveraging a metering capability, either pay-per-use or charge-per-use basis, to provide transparency for both the service provider and the consumer of the cloud.

3.3.4 *Service and Deployment Models*

With cloud computing being adopted at an increasing pace by both big and small companies worldwide, many firms have to make the decision what services they would like. Although currently the cloud community firmly advocates the Everything as a Service (EaaS) model for all companies, the cloud servicing providers still offer their products via different service models (Figure 3.2) and deployment models.

3.3.5 *Service Models*

3.3.5.1 *Infrastructure as a service*

The most powerful service that cloud computing firms provide is the IaaS. NIST defines IaaS as follows:

> "The capability provided to the consumer to provision processing, storage, networks and other fundamental computing resources where the consumer is able to deploy and run arbitrary software, which can include operating systems and applications."

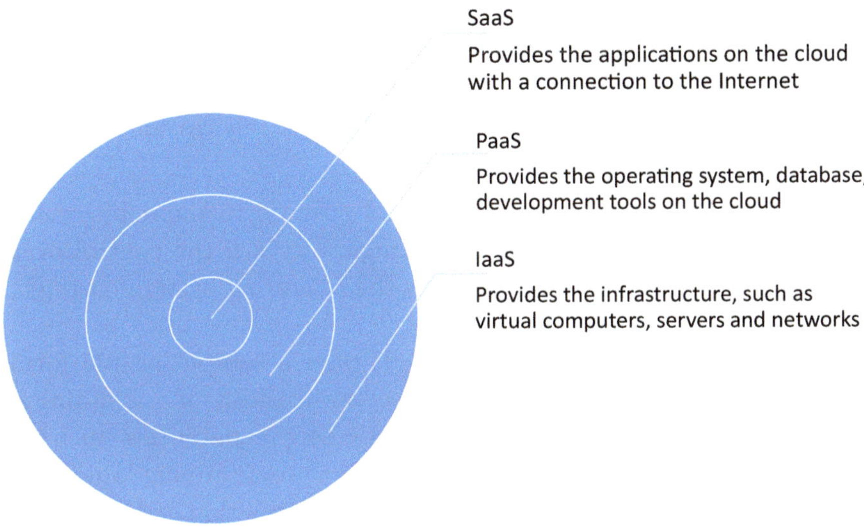

Figure 3.2: Summary of Service Models

Essentially, the consumer has control all the way down to the operating systems, storage and deployed applications.

In the IaaS, virtual computers are hosted by a hypervisor (a software, hardware or firmware that runs virtual computers) that provides consumers with a computing infrastructure such as data partitioning, backups or physical computing resources. To put this simply, what IaaS essentially does for consumers is to provide a virtual computer online where you can install an operating system (such as Windows) after which you can use it as you wish. This is beneficial for users that want to access the powerful performance of a computer but without having the hardware infrastructure to support it, instead doing it over a cloud server. IaaS clouds also provide add on resources such as firewalls, load balancers and IP addresses, along with software bundles.

3.3.5.2 *Platform as a service*

One level up from IaaS is the PaaS, where cloud service providers provide consumers the platform on which they can host their applications. NIST defines PaaS as follows:

> "The capability provided to the consumer to deploy onto the cloud infrastructure consumer-created or acquired applications created using programming languages, libraries, services, and tools supported by the provider."

In PaaS, the consumers have control only up to the deployment of applications and configuration settings, but lack the control over the underlying infrastructure such as the operating system, storage, networks, or servers.

In PaaS, cloud computing providers offer a computing platform where users can use it without the complications of maintaining the hardware such as hard drives, processor capabilities such as memory, and buying the software to support programming languages to develop their applications. The benefit to users that use the PaaS services is the removal of the headache of having the right hardware and software for certain power-heavy programs and applications, essentially handing over the maintenance and capabilities to the cloud service provider. Additionally, this platform can be accessed anywhere, essentially having a "Windows" on hand at any point of time with the Internet.

3.3.5.3 *Software as a service*

At the top of the layers and the most common form of cloud computing service is SaaS. NIST defines SaaS as follows:

> "The capability provided to the consumer is to use the provider's applications running on a cloud infrastructure."

In SaaS, the consumer has no control over the hardware, midware or software, instead just accessing an application online that does not require them to have the program on their own devices. Essentially, what this means is that all end consumers are able to access a particular set of applications from wherever they are, through whatever device at any time through a web-browser or email. Popular examples of this include Microsoft Office Online and Google Documents, which allows for real-time access, editing and saving.

Table 3.1: Types of Service Models and the Level of Control

	IaaS	PaaS	SaaS
Control of Infrastructure/ Server	Vendor	Vendor	Vendor
Control of Operating System/ Database	Consumer	Vendor	Vendor
Control of Applications/Software	Consumer	Consumer	Vendor

The benefit of SaaS for companies comes in the form of not having to buy several copies of the program/software, and have a readily available online program that they can pay per user instead of having a physical person order the program themselves to get a license. Additionally, collaboration efforts with the right applications become much more advantageous.

The traditional Service models are summarized in Table 3.1, where the level of control and owner of the various levels in a cloud computing process is shown.

3.3.5.4 *Other types of service models*

Besides the three standard models of cloud services, more and more models are popping up to cater to the needs of consumers, and a new model catered to mobile app developers is the Mobile backend as a Service (mBaaS).

In mBaaS, application developers are given the ability to link their applications to the cloud for storage and user management. Common uses are seen in mobile gaming apps, which use popular mBaaS platforms such as Google Play to manage the users and the data of the accounts they have on a game through a cloud server. Additionally, cloud backups for a consumer's gallery, messages, and social media data are all through third-party cloud mBaaS providers.

3.3.6 *Deployment Models*

While there is a multitude of service models used for consumers to choose, we now need to look into how cloud providers deploy these clouds through several models, which is summarized in Figure 3.3.

3.3.6.1 *Private cloud*

The private cloud currently is the most frequented deployment model adopted by both big and small companies worldwide. NIST defines the private cloud as follows:

> "The cloud infrastructure is provisioned for exclusive use by a single organization comprising multiple consumers..."

The private cloud serves only one organization, and anyone that does not belong to that organization does not have access to this cloud. An example would be a cloud drive that employees of a company go to in order to retrieve files that their colleagues have created, edited or pasted into the cloud. The cloud may be owned, managed and operated by the organization itself, or an external

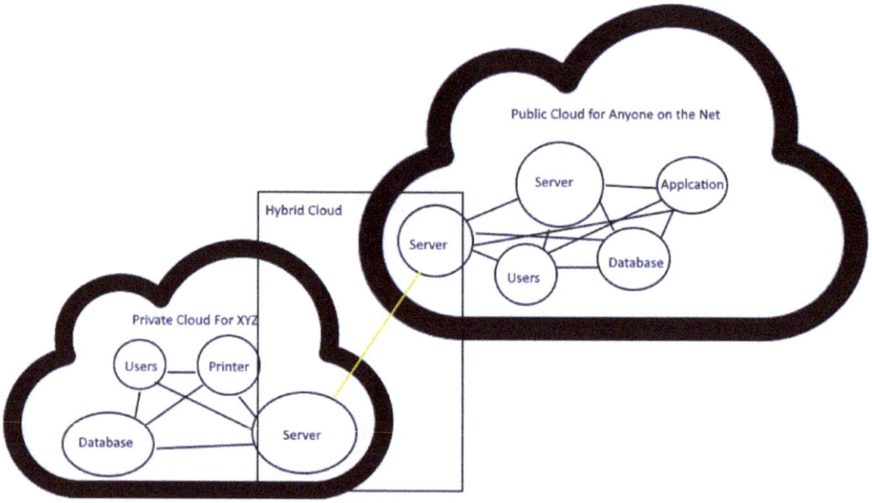

Figure 3.3: Summary of Cloud Deployment Models

third-party vendor, or a combination of both, and the physical servers may exist within the company's building or off-site.

3.3.6.2 *Public cloud*

The public cloud differs from the private cloud in the sense that the service that the cloud computing provider offers is available to the public. NIST's definition of the public cloud is as follows:

> "The cloud infrastructure is provisioned for open use by the general public."

While corporations may still utilize the public cloud and restrict access by passwords or user accounts, the public still shares the cloud space with them, hence limiting the security considerations for these services. The difference lies in whether the public is even able to access that server in the first place, as in private clouds, even with the directory or Uniform Resource Locator (URL), consumers outside of the organization would not be able to land in that page. Another stark difference is that the servers for public clouds are always at the premises of the service provider. Popular examples are Dropbox, Google Drive, and Amazon Web Services.

3.3.6.3 *Hybrid cloud*

The hybrid cloud is a model of deployment where organizations choose to link some part of their private clouds to the public cloud. NIST defines it as follows:

> "The cloud infrastructure is a composition of two or more distinct cloud infrastructures (private, community, or public) that remain unique entities, but are bound together by standardized or proprietary technology that enables data and application portability."

While most organizations have a private cloud for their employees, linking a part of the cloud to the public clouds enable them to achieve several benefits that using one or the other may not entail them to. One common use of hybrid clouds is when sensitive client data is stored within the private clouds, but links a portion of it to a business intelligence application that can be found on a public cloud (e.g., Dashboards) to utilize these programs and be able to share these with the appropriate clients.

Another common scenario of organizations opting for the hybrid cloud are IT firms who use the public cloud as a booster for their private clouds, in terms of computing capacity or data storage. This gives organizations the flexibility to limit the private cloud's capabilities to a certain level, and when more cloud computing resources is needed, they can "burst" into public clouds to lend their server's capacity in times of heavy usage.

3.3.6.4 *Other deployment models*

The three deployment models talked about in the previous section makes up the majority of cloud deployment in the modern business world. However, there are some specialized but less common deployment models as follows:

- **Community Cloud:** The community cloud is a semi-private public cloud that is shared only among several organizations that have a common cause or concern. Common examples of community clouds can be found in governmental agencies, or in the healthcare sector. One use of the community cloud would be among the different hospitals in a certain country, where they are able to access each other's patient records so as to facilitate easier processing of the patients that visit different hospitals.
- **Distributed Cloud:** Distributed cloud is a cloud computing platform that can be assembled from several machines, servers, or volunteers. In essence, distributed cloud requires several people or organizations to volunteer their computing resources to run the cloud, which is also known as peer-to-peer clouds. A danger of such a cloud is the volatility of the providers/volunteers, as they are not bound to any contract or organization that can be held responsible for a cloud outage. However, this cloud system has risen in popularity recently, with companies now leveraging on the sharing economy model by providing a platform where consumers can either volunteer their computer's resources or hire a

cloud resource, with people volunteering their computer's resources rewarded with an incentive.

- **Big Data and High-Performance Clouds:** With the amounts of data that companies collect on a daily basis, and the amount of individuals utilizing the cloud resulting in even more big data, big data clouds have become a common cloud deployment model by many data-oriented companies. With dedicated single servers serving an entire company, big data clouds have risen in popularity, and many of these clouds coincide with high-performance clouds, where the cloud computing service providers equip their cloud infrastructure with top-of-the-line computing power and memory that enables for the processing of huge amounts of data and multiple tasks together.

3.3.7 *Advantages of Cloud Computing*

The advent of cloud computing is no surprise, and several benefits have been put forward as the driving force of cloud computing for companies in the near future:

- **Resourcing:** Organizations would not need to deploy manpower resources such as technical support staff and physical hardware such as bulky servers that need a huge environment-controlled area to store them as well, hence releasing space for other functions. Additionally, their employees can focus less time fixing technical issues or learning how to set up their computers, and spend more time on the actual work of the company.
- **Costs:** Costs are relatively cheaper using cloud technology as it eliminates the need for buying and maintaining the hardware, midware, and software. Additionally, cloud computing leverages on economies of scale by assigning and re-assigning resources according to usage levels.
- **Maintenance:** Maintaining high-capacity servers and hardware requires technical experts, a controlled environment as well as regular updating. With cloud computing, companies benefit from the service provider taking on full responsibility in maintaining these servers.

- **Reliability:** Cloud computing harnesses the power of shared resources, hence while a singular workstation may choose to default and shut down, most cloud computing services have a back-up server that is ready to step in.
- **Scalability:** The word used by many experts in describing cloud computing is "unlimited". Cloud computing leverages the power of elastic resources to enable any user to get the amount of computing capacity needed at any point of time.
- **Mobility Working:** In today's globalizing environment, many companies offer employees the flexibility to work away from the office. Cloud computing supports this by allowing employees access to an infrastructure, platform, or service that can be accessed from heterogeneous devices at any place and time with an internet connection.

3.3.8 *Disadvantages and Limitations*

While the benefits are apparent, cloud computing does come with disadvantages and limitations that need to be taken into consideration before choosing a cloud service:

- **Modification:** Most of the time, cloud computing providers have a certain level of standardization, as it is down to the provider's capabilities and resources. By leveraging on economies of scale, cloud computing services are able to save companies the cost by also limiting the amount of customization one can make.
- **Downtime:** The worry for many companies that utilize cloud computing is what happens when the service provider or Internet provider has an outage? The answer is what you would expect it to be, the business has to suspend temporarily. Since cloud computing relies entirely on Internet connectivity and the server being up, an outage would result in a cripple of a company's major functions.
- **Security:** Cloud computing comes with the risk that anybody who is able to hack into the service provider's server will be able to gain access to several companies' data through one successful attempt, which is known as "hyperjacking" (Huang *et al.*, 2012).

- **Ownership:** When someone puts a data on the cloud, does that data belong to the cloud service provider as well? Terms surrounding the use of cloud computing have historically been vague and are a grey area in many legal systems around the world.

3.3.9 *Conclusion*

Although it is clear that there are multiple benefits as well as disadvantages to cloud computing, it is here to stay, as seen by the projected value of the industry and the amount of revenue that will go through cloud technology. The impact that cloud computing will have on businesses is no doubt a net positive, but managing the service carefully will be key in getting the most out of the exciting future of cloud.

The impact of Artificial Intelligence and Cloud Computing in the near future is clearly a game-changer that will bring about multiple changes at organizations across several industries. However, the downside of the ever-improving and widespread adoption of such technologies always come with new risks and security concerns, especially since these technologies essentially play a crucial role in the operation of businesses when adopted.

The benefits of these technologies, however indicate that companies would be wise not to ignore them based purely on the risk of security and concerns. Instead, firms need to take the first step into learning about how to integrate these technologies into their businesses by understanding their uses and benefits. The key in mitigating the security risks is proper planning for fail-safe situations, as well as gradual integration of these technologies rather than just dumping their entire operations onto AI and the cloud. With the world adopting these technologies at a phenomenal pace, companies would be left behind if they do not start soon.

References

Ayon, D. (2016). Machine Learning Algorithms: A Review. *International Journal of Computer Science and Information Technologies*, 7(3), 1174–1179.

Bernard, M. (2018). How Is AI Used In Healthcare — 5 Powerful Real-World Examples That Show The Latest Advances. [online] Forbes. Available at:

https://www.forbes.com/sites/bernardmarr/2018/07/27/how-is-ai-used-in-healthcare-5-powerful-real-world-examples-that-show-the-latest-advances/#4f084b655dfb. [Accessed 17 November 2018].

Brian, K., Matt, C., & Richard, F. (2018). *10 Promising AI Applications in Health Care*. [online] Harvard Business Review. Available at: https://hbr.org/2018/05/10-promising-ai-applications-in-health-care. [Accessed 17 November 2018].

Daniel, F. (2018). Machine Learning in Finance — Present and Future Applications. [online] Available at: https://www.techemergence.com/machine-learning-in-finance. [Accessed 14 November 2018].

Daniel, F. (2018). Artificial Intelligence in Retail — 10 Present and Future Use Cases. [online] TechEmergence. Available at: https://www.techemergence.com/artificial-intelligence-retail/. [Accessed 17 November 2018].

David, M. (2015). Support Vector Machines — The Interface to LIBSVM in Package e1071.

Deborah, W. (2018). Four Ways AI Is Leveling the Retail Playing Field Versus Amazon. [online] Forbes. Available at: https://www.forbes.com/sites/deborahweinswig/2018/07/28/four-ways-ai-helps-retailers-compete-with-amazon/#11469f2219cc. [Accessed 18 November 2018].

Deena, Z. (2018). The 3 Most Valuable Applications of AI in Health Care. [online] VentureBeat. Available at: https://venturebeat.com/2018/04/22/the-3-most-valuable-applications-of-ai-in-health-care/. [Accessed 17 November 2018].

Gartner (2010). Gartner Says Worldwide Enterprise IT Spending to Reach $2.5 Trillion in 2011. Retrieved from https://www.gartner.com/newsroom/id/1453519

Gartner (2017). Gartner Forecasts Worldwide Public Cloud Revenue to Grow 17.3 Percent in 2019. Retrieved from https://www.gartner.com/en/newsroom/press-releases/2017-10-12-gartner-forecasts-worldwide-public-cloud-services-revenue-to-reach-260-billion-in-2017

Huang, Y. L., Chen, B., Shih, M. W., & Lai, C. Y. (2012). Security impacts of virtualization on a network testbed. In *IEEE Sixth International Conference on Software Security and Reliability (SERE), 2012*. IEEE, pp. 71–77.

Insights Team (2018). How Artificial Intelligence and Machine Learning Are Revolutionizing Logistics, Supply Chain and Transportation. [online] Forbes. Available at: https://www.forbes.com/sites/insights-penske/2018/09/04/how-artificial-intelligence-and-machine-learning-are-revolutionizing-logistics-supply-chain-and-transportation/#57a8025958f5 [Accessed 19 November 2018].

Intelegain Team (2018). Role of Artificial Intelligence in Payment Services. [online] Available at: https://www.intelegain.com/ai-in-payment-services/. [Accessed 13 November 2018].

International Data Corporation (IDC) (2017). The Salesforce Economy Forecast: 3.3 Million New Jobs and $859 Billion New Business Revenue to Be Created from 2016 to 2022. Retrieved from https://www.salesforce.com/content/dam/web/en_us/www/documents/white-papers/idc-study-salesforce-economy.pdf

Jennifer, B. (2018). Top 12 Ways Artificial Intelligence Will Impact Healthcare. [online] HealthITAnalytics. Available at: https://healthitanalytics.com/news/top-12-ways-artificial-intelligence-will-impact-healthcare. [Accessed 17 November 2018].

Meyer, D., & Wien, F. T. (2012). Support Vector Machines. *R News*, 1(3), 23–26.

Meyer, D. (2012). Support Vector Machines. *R-News*, 1(3), 9.2001.

National Institute of Standards and Technology (NIST) (2011). The NIST Definition of Cloud Computing. Retrieved from http://faculty.winthrop.edu/domanm/csci411/Handouts/NIST.pdf

Richter, F. (2019). Cloud Business Drives Amazon's Profits. Retrieved from https://www.statista.com/chart/9174/amazon-operating-profit/

Samantha, D. (2018). How AI Is Changing the Retail Industry. [online] business.com. Available at: https://www.business.com/articles/artificial-intelligence-changing-retail/. [Accessed 18 November 2018].

Chapter 4

Introduction to Financial Technology

4.1 Introduction

In recent decades, the emergence of financial technologies (FinTech) — especially digital banking services — has re-shaped the financial industry around the globe. This chapter reviews these developments and shows how they have been influencing the structure of and competition in financial service industries and other business domains. The assessment of how FinTech and globalization affect businesses highlights the need for a deeper understanding of the FinTech ecosystem to mitigate risks and harness potential benefits.

4.2 What Is FinTech?

"FinTech" is an abbreviation for the phase *financial technology*, which started gaining traction after the 2007–2008 global financial crisis (GFC). Since technology plays an important role in the banking business, people in the technology industry start seeing their opportunities to disrupt the current state of financial services. Technology can help to improve the current service level provided by banks, by moving away from their current support and operation role to interacting and reaching out to service the customers directly. They see themselves as "disruptors", disrupting the current landscape of banks being the sole provider of financial services to them, and being a viable or even better provider of financial services. The financial innovations and the amount of investment capital channel into these FinTech startups are so much that stalwarts of the banking industries are wary of the threats they are posing to their business. Jamie Dimon, Chairman and CEO of J.P. Morgan, mentioned in his 2014 letter to his shareholders[1]:

> "**Silicon Valley is coming.** There are hundreds of startups with a lot of brains and money working on various alternatives to traditional banking.... They are very good at reducing the 'pain points' in that they can make loans in minutes, which might take banks weeks. We are going to work hard to make our services as seamless and competitive as theirs. And we also are completely comfortable with partnering where it makes sense."
>
> — Jamie Dimon, Chairman & CEO of J.P. Morgan Chase

4.2.1 *Post-GFC Effects*

Tech entrepreneurs are interested in being financial providers because the current environment is very much in their favor. During the GFC, taxpayers' money has been used to bail out the irresponsible business practices of bankers. Post GFC, financial reforms have been instituted, which while stabilizing banking and reducing risk profiles also had the effect of limiting access to capital for businesses. The general sentiment of the banking industry is therefore of a negative note. However, public trust in technology is sky-high[2]; hence, an alternative provider of financial services is very much welcomed by consumers and business owners.

4.2.2 *E-commerce*

With the proliferation of e-commerce, online users, especially millennials are experienced with buying and selling items/services online and most of them are comfortable with the idea of parting

[1]JPM Chairman and CEO letter to shareholders 2014, http://files.shareholder.com/downloads/ONE/3844439630x0x820077/8af78e45-1d81-4363-931c-439d04312ebc/JPMC-AR2014-LetterToShareholders.pdf.

[2]Political risk stalks booming Silicon Valley, https://www.ft.com/content/2690a2d8-148f-11e7-80f4-13e067d5072c?mhq5j=e2.

their money to a total stranger, before receiving the purchased items/services. Tech companies, such as Amazon[3] and Alibaba started out as e-commerce platforms, and gradually, they started providing simple forms of payment/banking services to their users and merchants, such as the provision of escrow services. Providing more banking services to their existing users should be a logical extension of their existing business models, since they are already familiar with their spending habits and cash flow situation. They have huge user bases and they are always looking for new opportunities to monetize their users. Providing banking services can diversify their existing range of services and add more revenue streams.[4] For example, Google has an 88% market share in search advertising and Facebook controls more than 70% of social media on mobile devices.[5] Including a buy button in their search results would allow them to get a share of the growing e-commerce pie.

4.2.3 *Global Financial Inclusion*

In 2010, World Economic Forum started the Global Financial Inclusion[6] project that looks into advancing financial inclusion to over 2 billion people worldwide and 200–250 million medium-to-small enterprises because neither pure cash systems nor the traditional banking model can effectively serve them. "Financial inclusion" refers to the provision of useful and affordable financial products/services that individuals and businesses need in the areas of transactions, payments, savings, credit and insurance. For technology groups, the ability to make use of technology to provide financial services to the underserviced fulfills the social mission in their business. It is only the use of technology in financial services that allows people with an

[3]Another industry Amazon plans to crush is small-business lending: Op-Ed, http://www.cnbc.com/2017/06/16/amazon-plans-to-crush-small-business-lending.html.

[4]Bank of Tech poses growing threat to traditional institutions, https://www.ft.com/content/1a862cd2-efd3-11e6-ba01-119a44939bb6.

[5]Silicon Valley has too much power, https://www.ft.com/content/60a49086-3714-11e7-bce4-9023f8c0fd2e.

[6]Promoting Global Financial Inclusion, https://www.weforum.org/projects/promoting-global-financial-inclusion.

income of US$1.25 per day to be serviced, for the use of technology greatly reduces the operating costs of onboarding a new customer.

According to Ernst & Young (EY)'s FinTech Adoption Index 2017, the average percentage of digitally active consumers using FinTech services reached 33% across the 20 mature and developing markets. Across the 20 markets, Chinese consumers have the highest adoption rate of 69%. FinTech represents disruptive innovation that alters the way financial service markets operate and has forced financial institutions to rethink their approach to the industry. FinTech adoption will continue to gain momentum.

4.3 How FinTech Is Changing the Consumers?

FinTech has been changing customer expectations and behaviors at a tremendous pace. Customers expect a seamless mobile experience. The industry is changing rapidly, and businesses should embrace the latest technologies and stay at the forefront of their markets.

There are many drivers to success:

- **Offer a previously paid-for service free of charge or at a cheaper rate:** In markets where customers are highly price-sensitive, FinTech firms can achieve significant new customer acquisition through leveraging on technology to reduce their costs and maintain this competitive advantage as they grow.
- **Provide a new type of value-adding service:** A FinTech firm could fulfil previously unmet customer needs. Consumers are now able to access data and information anytime and anywhere and are able to adjust their investment portfolio on the go.
- **Solve a problem for another business:** A FinTech firm could help a more established business to offer its customers better services, e.g., new credit-scoring services supported by data analytics.
- **Focus on previously underserved customer segments:** FinTech firms can offer a new product or expand an existing product to a new segment, e.g., FinTech Firms may apply a different set of metrics and risk appetites to designing new loans, enabling

a traditional retail bank to provide mortgages to a previously underserved population.

4.4 FinTech Applications

A majority of FinTech startups focus on their operations and businesses by reaching the customer directly, bypassing traditional banks and disrupting a sector ripe for innovation. The general areas of interest and investment in FinTech startups have mainly focused on the following areas.

4.4.1 *Mobile Payments*

Mobile payments refer to the third-party providers and/or developers of payment apps and platforms which allow their users to pay bills and transfer money. Some examples of mobile payments are Alipay in China, SingTel Dash in Singapore, and Venmo (a subsidiary of PayPal) in US.

4.4.2 *Crowdfunding and Peer-To-Peer (P2P) Financial Services*

Businesses can raise funds quickly and cheaply from people all over the world through crowdfunding platforms. It has democratized the process of sourcing for start-up capital and shortened the timeline.

P2P financial services refer to the online platforms/businesses which match people who need a financial service with others who need it in the reverse direction. An example of P2P financial service is P2P lending, where the platform will match borrowers with lenders. The business model of such platforms normally charges a fee for providing the matching services. Some examples of P2P lending are provided by EasiCredit in China, MooLahSense in Singapore, and Lending Club in the US.

4.4.3 *Quantitative Trading and Robo-advisory*

Quantitative trading and robo-advisory refers to the use of algorithms to carry out trading, portfolio selection and re-balancing. The huge amount of numerical data available in financial markets make it suitable to be subjected to analysis and optimization. The extensive use of computing power to carry out analysis makes it cheaper than using human analysts. FinTech startups often touted their low-cost base that translated to zero management charges for their investors.

The use of algorithms is no longer confined to quantitative trading. In the asset management industry, there are a number of automated wealth managers who are making use of computers and algorithms to offer sound financial advice to investors, especially to millennials. These automated wealth managers are called robo-advisors, because of the algorithms they deploy for asset management. With the pervasiveness of technology into our daily lives, millennials are naturally comfortable with investing their money as advised by their robo-advisors. These automated wealth management platforms allow investors to start investing with no minimum account balances. Because of the use of algorithms and technology, the cost of on-boarding a new investor is almost zero, allowing investors with small sums for investments to be serviced.

Some examples of robo-advisors are Bettlement and Wealthfront in the US. These robo-advisors allow an initial investment account with US$25 to start investing, unlike traditional investment advisors that need at least US$100,000, if not more.

4.4.4 *Cryptocurrencies and Blockchain*

Cryptocurrencies refer to digital currencies where encryption techniques and consensus algorithms are used to determine the monetary policy in the absence of a custodian, such as the central bank. The blockchain is the underlying technology that provides the security of the cryptocurrency. There are close to 850 cryptocurrencies available and some notable examples are Bitcoin, Ether, and Dash. Bitcoin is recognized as a commodity by the Commodity Futures Trading Commission in 2015, giving it a status just like oil and gold.

4.4.5 *RegTech*

Regulatory technology (RegTech) describes the use of technology in the context of regulatory monitoring, reporting, and compliance, particularly in the financial services industry. The innovations in

RegTech have focused on making data and new analytical methods easier for financial institutions to comply with regulatory requirements. Streamlining and/or automation of processes allow for more effective and efficient financial risk identification and regulatory compliance. For example, tools like real-time risk analysis dashboard can help institutions spot fraud more quickly. Commercial banks and regulators have been the main target audience for RegTech. As the industry emerges over time, the applications of RegTech may extend to retail banking.

4.4.6 *InsurTech*

Insurance Technology (InsurTech) is a subset of FinTech which relates to the use of technologies and data to simplify and improve the efficiency of the insurance industry. InsurTech offers customized insurance solutions to life risks using data analytics, sensors, wearables and location-based mobile phone data, with a strong focus on data integration, machine learning, and on-demand services. Traditional insurance firms are facing increasing competition due to the emergence of a variety of innovative InsurTech solutions. Examples of InsurTech activities are indicated in Figure 4.1.

There are many other areas of FinTech, such as FinTech startups that deal with risk management or the provision of data analytical skills/expertise/products for the operational aspects of banking. Alternative credit scoring and assessment, based on alternative data

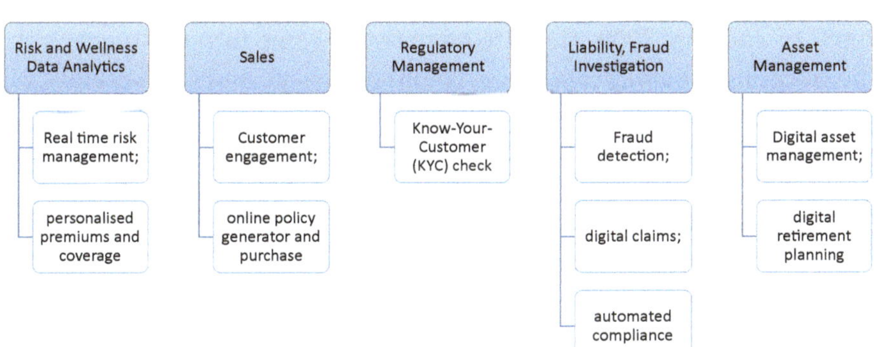

Figure 4.1: Examples of InsurTech Activities.

such as social interactions and future prospects, is also an actively pursued area in the area of FinTech startups.

FinTech has the advantage of lowering the cost and being an enabler for new and profitable business models. FinTech applications like crowdfunding, mobile payments, and payment transfer are revolutionizing the way small businesses startup, accept payments, and go global, and they are making it easier than ever to start and run a business.

4.5 Cryptocurrencies and Blockchain

One of the most exciting applications of FinTech is the cryptocurrencies. A cryptocurrency is a decentralized digital currency using encryption — the process of converting data into code — to generate units of currency and validate transactions independent of a central bank or government. Bitcoin and Ether are the most common cryptocurrencies. Cryptocurrencies are typically open access and have great potential to automate financial services with improved security and transparency.

4.5.1 *Bitcoin*

Bitcoin is the first and has been one of the most prominent cryptocurrencies used by traders in the crypto field. Bitcoin came into the scene with the publication of the paper *Bitcoin: A Peer-To-Peer Electronic Cash System* by Satoshi Nakamoto in November 2008. In January 2009, the bitcoin network came into existence with the release of the first open-source bitcoin client and the issuance of the first bitcoin to Satoshi Nakamoto.

Since then, the development of bitcoin was fast and furious. In June 2011, Wikileaks and other organizations began to accept bitcoins for donations and the bitcoin reached parity with the US dollar in February/April 2011. In October 2012, more than 1,000 merchants were accepting bitcoins under their payment processing service, and by 2013, 160,000 merchants were accepting bitcoins for payments. From 2015 onward, the US recognized bitcoin as a commodity and Japan recognized it as a currency in 2016. In March

2017, the price of 1 bitcoin surpassed the spot price of an ounce of gold for the first time.

Today, bitcoin can be traded by going to the few bitcoin exchanges that provide the selling and buying of the commodity. As of January 2020, the market capitalization of bitcoin is about US$177.2 billion (see Figure 4.2).[7] After bitcoins are purchased, they are typically kept in a Bitcoin Wallet in your mobile phone or computer and secured with your personal key.

Bitcoin is a digital currency that is not printed by a central bank, thereby making it unaccountable to any political powers or population. The value of bitcoin is derived from the supply and demand economics and is free from political monetary policy, i.e., the devaluation of a country's currency in order to curtail inflation or to pay for national debts. Bitcoin is created digitally, by a community of developers, users, investors, and miners. Anyone is welcome to join the bitcoin network either as a developer, investor, user, or

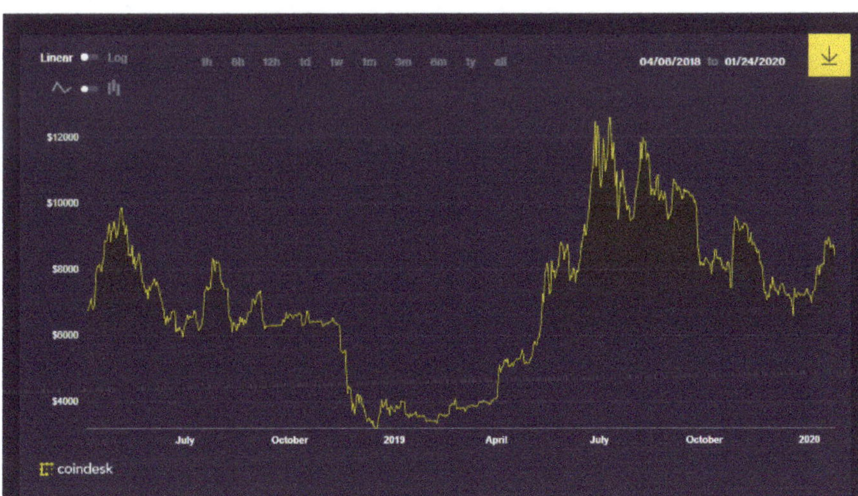

Figure 4.2: Bitcoin Price.
Source: https://www.coindesk.com/price/bitcoin.

[7]CryptoCurrency Market Capitalizations, https://coinmarketcap.com/currencies/bitcoin/.

miner. There are no administrators in the bitcoin network to vet a person's credibility. Bitcoin is therefore a permission-less blockchain network. Despite the absence of administrators to safeguard the bitcoin network, the network is safe from hackers due to the use of cryptography and the blockchain technology.

As the bitcoin network is designed to run without any administrators, all rules and protocol regarding how the network works are all coded into the network by developers. It is crucial that bitcoin participants follow rules so as to maintain a state of consistency across the network, thus guaranteeing security of the network.

Within the bitcoin protocol, there are necessary codes which govern

- mining, i.e., the creation of new units of bitcoin;
- processing of transactions.

Bitcoin Core houses the protocol within the bitcoin network and it was first developed by Satoshi Nakamoto himself. Subsequently, it was maintained by a wide group of independent developers, which is referred to as a community. The current lead developer for Bitcoin Core is Wladimir J. van der Laan and the entire code for Bitcoin Core is available for anyone to download or modify.[8] The developments are based on proposals and the public are welcome to vet the suggestions. Developmental work will only be carried out if there is sufficient support for the recommendations. The general approach for carrying out improvements in the bitcoin protocol is to gradually add, rather than rapidly add capabilities.

It should be noted that the governance of the bitcoin protocol is not done by the developers. It is actually the users who will decide which (version of) bitcoin protocol to use, when they download the version of bitcoin protocol to run on their computers. Developers are best understood as tool providers, since they code up the protocol. Their influence is limited to offering people software they can use to connect to the bitcoin network if they want to. The status as developers does not grant them any special privilege in bitcoin governance.

[8]Team | Bitcoin Core, https://bitcoincore.org/en/team/.

4.5.2 *Blockchain Technology*

Blockchain is the underlying technology that allows bitcoin to prosper as a viable currency. A blockchain is actually a database, or in bitcoin terms, a ledger, of all bitcoin transactions that have ever occurred.

The blockchain for bitcoin has the following characteristics, namely

- distributed, i.e., every node in the bitcoin network has a copy of this ledger,
- decentralized, i.e., there isn't any administrator for this ledger,
- transactions which have been accepted and confirmed by the blockchain cannot be changed,
- transactions in the blockchain are chronologically ordered.

There are three parts to a blockchain — the transactions, the block and finally the chain of all the blocks, i.e., the blockchain.

4.5.2.1 *Transactions*

A transaction in a bitcoin network is an agreement between two or more parties on how a sum of bitcoins is to be used. A transaction must have input(s), amount, and output(s) defined. Once the transaction is set up and announced to all the nodes in the network, the nodes will ensure that it is a valid transaction. A valid transaction is defined as a transaction where

1. there is no double-spending, i.e., all the inputs to the transactions have not been previously spent, and this can be confirmed by checking through the blockchain, since every node has a copy of all the transactions that have ever occurred;
2. the amount that is referenced in all the inputs has been properly accounted for;
3. a user who created the transaction is the rightful owner of the bitcoins; this is verified by checking the digital signature of the owner of bitcoins.

If a transaction does not satisfy all of the requirements, it will be dropped by the node of the network and will not be propagated onward to other nodes in the network. Once the transaction is confirmed to be valid, it will be considered to be included into the next block of the blockchain.

4.5.2.2 *Blocks*

A miner in the bitcoin network will amass a list of valid transactions and will start to mine for a new block, which consists of the following items:

- the list of the hash of the most recent valid transactions — the list of all valid transactions will be hashed together into a Merkle tree structure;
- the hash of the last block in the blockchain;
- a nonce.

These are three items that are used to generate the hash of the new block. A new block is generated via the bitcoin mining process, i.e., when the nonce is changed such that it can solve the computationally difficult cryptographic puzzle to the required difficulty level. The mining process incentivizes miners to validate transactions, as they are rewarded with new bitcoins for checking the validity of all new transactions in the network.

As the hash of the new block is dependent on the list of the most recent transactions, changing any transaction details will instantly invalidate the hash of the new block. Therefore, once a transaction is included into the block, its details cannot be changed. A valid transaction that is included into a block is said to have been "confirmed".

4.5.2.3 *Chain of the blocks*

Once a block is found, it is included into the blockchain. New blocks will be discovered at an average rate of 10 minutes and the older the block, i.e., the earlier a block is cascaded into the chain, the harder it will be for a malicious user to change the details of a transaction. Hence, transactions in an older block are considered to be more secure, and it is recommended that these transactions receive six block confirmations before they are considered to be irreversible.

There are occasions in the bitcoin network that two blocks are discovered simultaneously, resulting in a "fork" of the blockchain. This is not allowed and it will be resolved when the next block is discovered. The bitcoin protocol states that a new block must chain to the longest branch of the blockchain. Subsequent blocks must chain to the longer branch of the blockchain and the forked block will become an orphan block and all the transactions in the orphan block will be included in the subsequent newly discovered block.

A large network is bound to suffer from network latency effects and so will the bitcoin network. Latency effects in a bitcoin network will result in different nodes in the bitcoin networks receiving the latest transactions at different times. Despite the latency problem, the bitcoin network is able to prosper due to the design of the blockchain.

The reason why bitcoin has been able to take off as a viable cryptocurrency is because of the security of the blockchain, the health of the mining ecosystem, and the value of the bitcoin currency. The blockchain has to be secure in order for the bitcoin to prosper and for the blockchain to be secure, a malicious user cannot overwhelm the consensus process, i.e., taking control of >51% of the bitcoin network. To have a healthy mining ecosystem, we need lots of (individual) miners putting in a lot of computing power to solve the puzzle. Miners will only be incentivized to join the mining process when the exchange rate of bitcoins is high.

There will only be a high and stable value of the currency when users in general trust the security of the blockchain, i.e., the network will not be overwhelmed by an attacker anytime. It is these three interlocking interdependences that enable the bitcoin to take off as a viable cryptocurrency, despite the long history of failed attempts to create digital currency schemes.

4.5.3 *Blockchain Implications*

The bitcoin network is interesting, because it is a network that is self-regulating and is able to make decisions by itself, without any human intervention. All the decision-making protocols have been coded into the bitcoin protocol. Anyone is welcomed into the network to make money, i.e., to mine for bitcoin or to spend money, i.e., create a bitcoin transaction and the bitcoin network does not make any assumption on whether a new user is trustworthy or not. There is no concept of *trust* in the network. This is certainly not how we are operating in our current environment. Before onboarding you as a new customer, banks are required to do Know Your Customer (KYC) and the bank would not accept you as their deposit customer, if they think you are dealing with, say, fake currency. Similarly, from the customer's perspective, you would not go to a dodgy bank to deposit your hard-earned money, because you will not trust your money with them. Our current environment operates with the premise of trust. In bitcoin, the network does not care, because the consensus of the network is able to deal with malicious users. See Figure 4.3.

Because of the need of trust in our current business transactions, we often need financial/legal intermediaries in our business dealings. Consider the way we conduct business these days. We might need a bank to provide us with a document to say that we have the necessary financing to pay for our goods and we will need a lawyer to bind our supplier to us to provide the necessary goods for our business. The bank and the lawyer are the intermediaries for our business. They are the trust bearers and they do not come free. We incur costs in

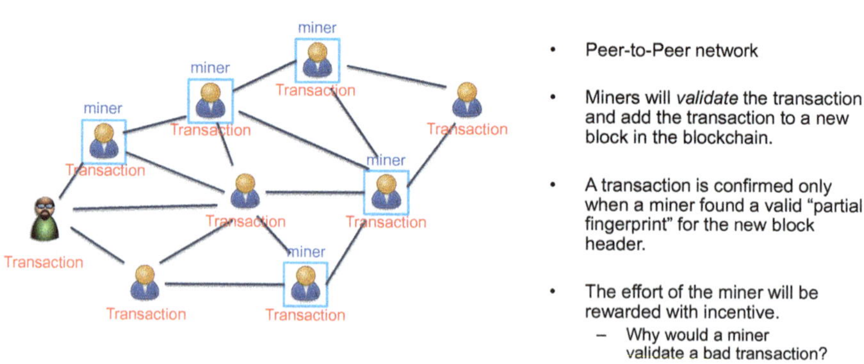

Figure 4.3: Blockchain's Provision of "Trust" — the [Economy] Part. *Source*: By authors.

engaging their services because without them, the potential losses could be greater, if the counter party of our dealing does not honor their part of the deal.

With blockchain, the way we conduct business will change. We can set up our legal agreement with our supplier, and both parties will sign digitally on the legal agreement. Instead of getting a lawyer to be a part of the agreement, i.e., a witness to the agreement, the legal contract can be a transaction in a blockchain. Once our business agreement gets into the blockchain, the business agreement cannot be changed, because of the immutable property of the blockchain. If both parties to the agreement want to change any terms in the legal agreement, they can again create another transaction which they both sign on and get it into the blockchain, with the previous legal agreement as an input to the new transaction. Again, there is no need to involve a lawyer to bear witness to these changes. The trust we have placed on the lawyer, as a fair witness of our business transaction is being replaced by the blockchain.

This is the new future that is visualized by proponents of the blockchain technology. They believe that with blockchain, "friction" or "pain points" in transactions can be significantly reduced. There is no need for intermediaries in our business transactions if every agreement on paper is moved onto the blockchain. However, whether this reality will come true will depend on the uptake of the blockchain technology worldwide.

4.5.4 *Applications of Blockchain Technology*

In the example discussed earlier, blockchain is used as a replacement of a lawyer as the fair witness of a business contract between two parties. The properties of the digital signature scheme with the immutable property of a transaction, once it gets into the blockchain, is what makes blockchain a suitable replacement for the lawyer. Blockchains have many properties that make it suitable for business uses and these properties are being exploited by technologists worldwide. In the following sub-section, we will look at a few interesting applications of blockchain technology.

4.5.5 *Distributed Ledger Technologies*

Recall that blockchain is essentially a distributed and decentralized database. There have been significant interests among central bankers and the banking community to make use of blockchain as the next-generation ledger. In the current banking systems, when someone transferred an amount from one bank to another bank, the first bank account must be debited and the second account must be credited with the same amount. If we have a distributed ledger using blockchain, every bank has the same copy of the distributed ledger. With the use of blockchain, there is no longer a need for each bank to do the required debiting and crediting. This will greatly simplify the reconciliation process that follows after every single transaction. Since transactions are immutable after they're included into a block, this makes the distributed ledgers highly secure and less vulnerable to attack from hackers. Banks are very interested in DLT, and R3CEV LLC[9] has been set up to look into the use of blockchain for such purposes.

From the central bankers' perspective, if all the banks are on the ledger system, it is very easy to trace the transaction details, should the need arise, for the purpose of investigation. There is no need for the central banks to wait while each bank is retrieving transaction details from their in-house system.

4.5.6 *Smart Contracts*

Smart contracts are snippets of code included in transactions, waiting to be executed when certain events occur. Smart contracts facilitate, verify, or enforce the negotiation or performance of a contract, thus making a contractual clause unnecessary. Smart contracts emulate the logic of contractual clauses and proponents of smart contracts claim that many kinds of contractual clauses may thus be made partially or fully self-executing, self-enforcing, or both. Smart contracts aim to provide security superior to traditional

[9] r3 webpage, http://www.r3.com/.

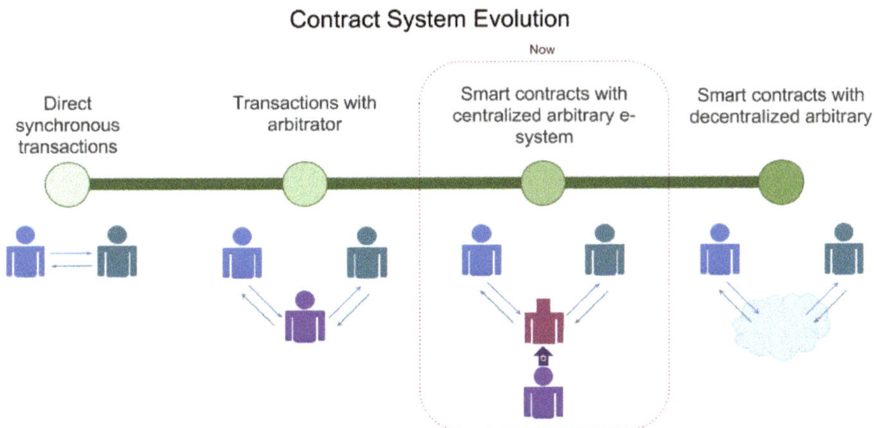

Figure 4.4: Contract System Evolution.
Source: Lee and Deng (2018).

contract law and to reduce other transaction costs associated with contracting.[10] See Figure 4.4.

Once the supplier's contract is placed in a blockchain and the blockchain is fed with data that the goods have been received, the smart contract will transfer the required payment of goods from the company to the supplier. All these are done with minimal human intervention. This will greatly reduce the paper work and delays involved in a typical goods-received-payment-due situation. Smart contracts are particularly suited for post-trade settlement, where there are lots of work and people involved — risk hedging, payment, settlement, collateral posting, and client's communication when a trade is done in banks.

4.5.7 *Initial Coin Offering*

Initial coin offering (ICO) is similar to initial public offering (IPO), but instead of handing over company ownership in the form of shares, a company doing an ICO sells a cryptocurrency token to

[10]Smart contract | Ethereum Documentation, http://ethdocs.org/en/latest/introduction/web3.html#smart-contracts.

the investors. Investors will pay ethereum (a cryptocurrency) to the company in exchange for the cryptocurrency tokens, which can be traded if the value of tokens increases or they can hold onto the tokens. These tokens do not give investors ownership of the company they had invested in but they allow early investors to the services that the company is offering. For example, Storj,[11] a decentralized storage solution issued tokens that can be exchanged for storage space on the platform. Storj's ICO raised US$30 million through the use of ICO.[12]

ICO is a new way that allows startups to get crowdfunded very quickly.[13] It allows startups to be funded by potential users, rather than venture capitalists. The popularity of ICOs is said to be responsible for making ethereum the next popular cryptocurrency after bitcoin.

4.6 From FinTech to Inclusion

Financial inclusion and impact investment are gradually viewed as mainstream activities. With an estimated 2 billion individuals and 200 million micro, small, and medium enterprises (MSMEs) in emerging economies considered as unserved or underserved by the formal financial system, financial inclusion has emerged as a critical driver to economic development. Many MSMEs and individuals transact exclusively in cash, with no secured way to savings and investments and limited access to credit beyond informal lenders and personal networks.

As more and more new technologies are introduced to the market, the boundaries of traditional business models are challenged. Using digital platforms rather than brick-and-mortar bank branches can dramatically reduce costs for service providers, providing financial

[11]Storj's website, https://storj.io/.

[12]The ICO is a revolutionary new way to get funded, and everyone wants in, http://mashable.com/2017/06/18/ico-explained/#XYbwmOI3A5qF.

[13]Ethereum unleashed the ICO craze, but it cannot handle its insane success, https://qz.com/1004892/the-bancor-ico-just-raised-153-million-on-ethereum-in-three-hours/.

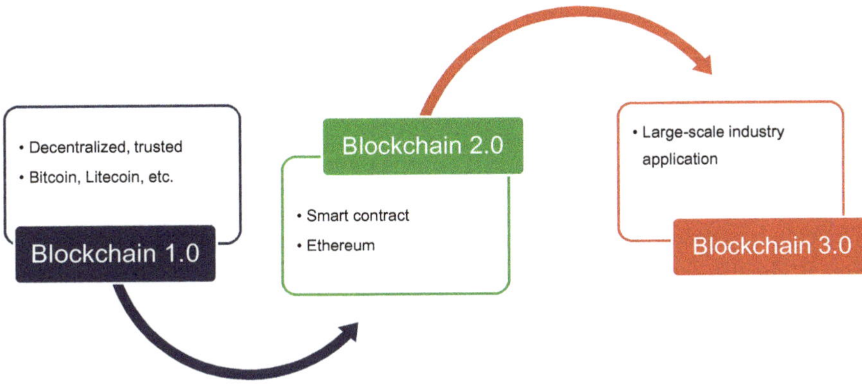

Figure 4.5: Blockchain Evolution.
Source: Lee and Deng (2018).

solutions to MSMEs and individuals at all income levels in rural areas. Digital financial services can also provide greater accountability, efficiency, and accessibility. See Figure 4.5.

Financial inclusion is the key to scalability and sustainability, as there are still more than 60% of the world population who are still unserved or underserved by the formal financial system and have been excluded from the economic and social system.

4.6.1 *FinTech Inclusion in ASEAN + China*

By 2015, ASEAN Economic Community (AEC) with its goal of regional economic integration, including a single market and production base, a highly competitive economic region, a region of fair economic development, and a region fully integrated into the global economy, is completed. AEC's areas of cooperation include human resources development; recognition of professional qualifications; closer consultation on macroeconomic and financial policies; trade-financing measures; enhanced infrastructure and communications connectivity; development of electronic transactions through e-ASEAN; integrating industries across the region to promote regional sourcing; and enhancing private sector involvement.

In ASEAN, consumers with low financial literacy rely on informal financial services. As the number of products and complexity increase, bad decisions and bad actions can hurt financially uneducated consumers. FinTech and especially blockchain can provide another avenue to protect consumers with transparency and big data collected and reported by the central banks. With the Belt and Road Initiative (BRI), ASEAN presents great potential for inclusive FinTech that promotes greater access to financial services. Inclusive FinTech has become an enabler for economic development and will propel the financial inclusion vision.

4.6.1.1 *Singapore*

Singapore is one of the financial hubs in the Asia-Pacific region. It has been ranked first in The World Bank's Ease of Doing Business study for the last 10 years. Also, it is regarded as one of the foremost examples of a smart nation with a wealth of initiatives to grow its FinTech start-up ecosystem.

With one-third of citizens aged between 25 and 34 years, Singapore has a rather high mobile penetration rate of 152%, standing as the country with the highest penetration of mobile broadband subscriptions per capita in the world. In addition, Singapore has an Internet penetration rate of 80%, which is the highest in Southeast Asia. All of the above conditions have provided a strong foundation for the emergence and growth of FinTech enterprises in Singapore.

Much is attributed to a visionary government policy to drive Singapore toward its positioning as the center in the Asia-Pacific region for startups playing their roles in the FinTech industry. The Singapore government has initiated numerous schemes and grants to help develop this potential sector.

In July 2015, the Monetary Authority of Singapore (MAS) announced to set up a new FinTech and Innovation Group (FTIG), which would be responsible for "regulatory policies and development strategies to facilitate the use of technology and innovation to better manage risks, enhance efficiency, and strengthen competitiveness in the financial sector."[14]

[14]http://www.mas.gov.sg/news-and-publications/media-releases/2015/mas-sets-up-new-fintech-and-innovation-group.aspx.

Furthermore, some government agencies, such as Singapore Economic Development Board (EDB), IMDA (Info-communications Media Development Authority) and Enterprise Singapore, have also taken a set of measures to build a favorable setting for technology startups. Such measures include grants, tax incentives, co-investment, conferences, etc.

4.6.1.2 *Project Ubin*

On 16 November 2016, MAS announced that it was partnering with R3 — a blockchain-inspired technology company and consortium of the world's largest financial institutions — on the production of a proof-of-concept (PoC) to conduct inter-bank payments facilitated by DLT. This endeavor, known as Project Ubin, is a digital cash-on-ledger project run in partnership between MAS and R3, with the participation of Bank of America, Merrill Lynch, Credit Suisse, DBS Bank, HSBC Bank, J.P. Morgan, Mitsubishi UFJ Financial Group, OCBC Bank, Singapore Exchange, United Overseas Bank, as well as BCS Information Systems as a technology provider.

The aim of Project Ubin is to evaluate the implications of having a tokenized form of the SGD on a DL and its potential benefits to Singapore's financial ecosystem. Being Singapore's central bank and financial regulatory authority, MAS acts as a settlement agent, operator, and overseer of payment, clearing and settlement system in Singapore that focus on safety and efficiency.

Many banks are working on FinTech and blockchain in Singapore. OCBC has been one of the most innovative advocates of blockchain among the banks in Singapore. The team at OCBC demonstrates its ability to use blockchain technology for cross-border transactions and remittances.

Given the enormous base of professionals in Singapore in the legal, audit, and corporate secretarial services and the ease of setting up companies and foundations, Singapore has been the favorite base for FinTech innovators. Almost all the successful ICOs are headquartered in Singapore with a foundation or business entity as contracting party for their projects and experiments.

4.6.1.3 *Indonesia*

By and large, Indonesia is the largest economy and state, with over 14,000 islands as an archipelago and has the fourth largest population in the world. Indonesia has a very structured financial inclusion framework that targets the very poor, working poor and near poor. They work through two channels: financial products and services (savings, credit, insurance, remittance, pension fund, mutual fund, and others), as well as public finance (subsidy, fiscal incentives, social welfare, cash transfer, health coverage "Jamkesmas", and others).

President Joko "Jokowi" Widodo rolled out the 14th economic reform package including a presidential decree on e-commerce roadmap. Similar to other ASEAN countries, funding for startups is a major pain point. The six solutions put forth by the Government were Crowdfunding, Micro Credit, Grants for Business Incubators, Universal Service Obligation (USO) Funds for digital-based SMEs, Angel Investment, and Seed Capital from Venture Capital. The USO, started in 2005 and halted in 2015, was originally provided for telco infrastructure in the rural areas, charged at 1.25% (0.75% in the beginning) of the telco companies' gross revenue.

According to Indonesia's FinTech association, there were more than 140 FinTech companies by the end of 2016. A majority of the companies were in the Payment (43%), Lending (17%), and Aggregator (13%) categories. Personal Financial Planning and Crowdfunding make up for around 8% of the startups. Indonesia's FinTech market is booming and grew by 78% from 2015 to 2016 with 20 foreign venture capital firms buying stakes in the local FinTech scene. There are already eight FinTech startups officially registered with the Financial Services Authority (OJK) by June 2017.

As Indonesia migrates from its old economy to a new emerging, young, and innocent ASEAN digital economy, telecom finance in Indonesia is already here. Three companies that have great implications for telecom-based financial inclusion in the future are Blanja as Telkom owning 60% and eBay 40%; Elevenia as XL Axiata owning 50% and SK Planet 50%; and Cipika, owned wholly by Indosat.

Other FinTech startups in cryptocurrency and blockchain are starting to grow. There are at least 11 cryptocurrency exchanges that one can have access to purchase cryptocurrencies in Indonesia. Asosiasi Blockchain Indonesia (the Blockchain Association of Indonesia, or idBitcoin) appears as a special form of association for all the bitcoin users in Indonesia with a vision to legalize the use of cryptocurrency and ensure it is used positively in the country.

4.6.1.4 *Malaysia*

Among the middle-income countries, Malaysia recorded the highest innovation performance. In this circumstance, there has been a strong drive to grow the FinTech sector in Malaysia. The high level of digital adoption and Internet penetration in Malaysia also makes it a potential FinTech market. Ranked fourth in global financial market development by the World Economic Forum (WEF), Malaysia has positioned itself as a future major financial center by its proximity to a large potential customer base of international financial institutions and end users. For instance, the global Islamic financial market, which is worth about US$2 trillion, is tightly connected to Malaysia.

Also, Malaysia's ambition to develop its FinTech sector is also a reflection of the growing need to reduce its economy's reliance on energy, given that the global oil prices would remain at a lower level within a period. So it is of little wonder that in 2015, Malaysia became the first nation in the Asia Pacific countries (APAC) to regulate for equity crowdfunding.[15]

Regarding regulation, in September 2015, Securities Commission Malaysia (SC) launched the "Alliance of FinTech Community" or "aFINity@SC" to impulse a network of FinTech stakeholders to accelerate growth and innovation in this sector.[16] At the beginning of this initiative, the SC would call for a gathering of FinTech stakeholders, such as entrepreneurs, investors, innovators, and relevant institutions, to discuss and set FinTech agenda together. Furthermore, FinTech startups could get advice and support to adapt

[15]http://www.austrade.gov.au/EventViewBookingDetails.aspx?Bck=Y&EventID=14508&m=0|0.

[16]http://www.sc.com.my/post_archive/sc-launches-afinitysc-at-world-capital-markets-symposium-2015/.

to the current regulatory environment. Effective discussions between enterprises and relevant authorities would be guaranteed to resolve regulatory and risk concerns.

In terms of capital, in 2003, the Malaysian government launched Cradle Fund to provide financial support for startups. About three years later, Malaysia Pre-Seed Fund was launched for the same goal. By 2014, these two funding programs had supported more than 800 startups in Malaysia.[17] At this stage, the emergence and growth of FinTech startups are continuously backed by the funding programs.

Maybank has cooperated with Malaysian Global Innovation and Creativity Centre (MaGIC) to boost FinTech startups not only in Malaysia but also in ASEAN.[18] MaGIC provides start-up services including legal advice, intellectual property consultations, company secretarial services and visa application via its own platform. Also, Maybank served as an advisory partner to MaGIC's Accelerator Programs and Academy to help the startups build viable propositions. This cooperation could boost social enterprise initiatives and also the development of FinTech startups in Malaysia and even the whole region.

4.6.1.5 *Thailand*

Thai financial regulators including The Securities and Exchange Commission (SEC) and the Bank of Thailand (BOT) are moving forward with supporting FinTech initiatives and equity crowdfunding specifically. The BOT, SEC, and Office of Insurance Commission are each looking to amend rules and provide channels for new market entrants within diverse areas of the financial industry. Simultaneously, the government has launched an SME private equity trust fund to help drive capital to SMEs. The fund has been capitalized with 2 billion baht.

---

[17]http://www.techtalks.ph/wp-content/uploads/2015/12/Philippine-Roadmap-for-Digital-Startups-FinalDraft_launch.pdf.

[18]http://FinTechranking.com/2016/01/06/maybank-magic-collaborate-to-boost-start-up-businesses-in-malaysia/.

To catch up with the new trend in FinTech development in other countries, the SEC is preparing to change the landscape of business's licenses within the capital markets based on four amendments. This includes the structure of the business licenses, the minimum capital requirement, the qualifications of the applicant, and other monitoring regulations where further details regarding these changes will be provided in the near future.

Even though FinTech is still at an early stage in Thailand, some companies have backed or launched initiatives to increase their interactions with the FinTech start-up ecosystem. The expectation of Thailand's FinTech startup is high with the transaction value in the FinTech market amounting to US$6,464.7 million in 2016. The transaction value is expected to show an annual growth rate (CAGR 2016–2020) of 19.99% resulting in the total amount of US$13,401.8 million in 2020. The market's largest segment is the segment "Digital Payments" with a total transaction value of US$6,440.9 million in 2016.

As the driver of innovation, the government attaches great importance to introducing foreign banks and developing FinTech. The banking sector is also involving itself actively to expand its reach and digital service. FinTech startups are also flourishing to take advantage of the trend.

FinTech Association of Thailand was set up with the aim to reduce the cost of financial transactions in Thailand, to increase accessibility to financial products and services for Thai people, and to support Thai FinTech startups to compete and expand to the global market. It intends to set up a National FinTech Sandbox and build an ecosystem for globalization.

4.6.1.6 *Vietnam*

The Central Bank or The State Bank of Vietnam has partnered with the World Bank Group to implement a national financial inclusion strategy since 2016. Digital finance is the focus with an aim of shifting government payments digitally and providing financial services to the rural, agriculture, and ethnic minorities. Financial literacy and consumer protection are keys to financial inclusion. Only one-third of the population has access to formal banking according to Global Findex database.

Among the ASEAN-4 which joined later, Vietnam since its *doi moi* has caught up almost as fast as and as well as some in ASEAN-6. FinTech startups in Vietnam are among the most dynamic in synchrony, given its strong entrepreneurial base and environment. Bitcoin, crowdfunding, mobile phones, and all other features of FinTech are among its many applications with a relatively young cohort of millennials. The country is as large and diverse for the growing expansion of innovative financial services.

Vietnam has a good number of cryptocurrency and blockchain companies, and in terms of developers, it has a good base that many have tapped from the region. Given its large population and underdeveloped financial inclusion space with qualified technology professionals, the potential is understated.

4.6.1.7 *Philippines*

The Office of the President issued Executive Order No. 208 on 2 June 2016 to institutionalize the Financial Inclusion Steering Committee (FISC), the governing body that provides strategic direction, guidance, and oversight of the National Strategy for Financial Inclusion (NSFI). The aims are to align various financial inclusion policies and programs, to ensure effective monitoring of the progress of initiatives, and to collaborate with public and private sector stakeholders on various agenda that fall under their focus areas.

Besides the interbank account-to-account fund transfer system enabled by the Electronic Payment and Clearing System, a new interoperability system allows consumers to pay anyone with a bank account or e-money account electronically.

In the Philippines, the use of microbanking offices (MBOs) is important to serve the rural areas. There were 540 MBOs operating in 338 local government units in 2015, of which 66 cities and municipalities were being served by MBOs alone. Out of 652 banks, 39 banks are using non-full-blown bank MBOs. MBOs have been growing at an annual rate of 22% since 2011.

With both Spanish and American colonization experiences, the Philippines can reach its potential economic growth and FinTech development when the government picks up the speed with reforms and job creation. The Filipino returnees will bring along their experiences working overseas, from the Middle East to other ASEAN states, as well as Hong Kong, together with business and social networks. FinTech definitely is part and parcel of that package to jumpstart the Philippines.

4.6.1.8 *Myanmar*

The financial scene in Myanmar provides ample opportunities for payments, savings, credit, and insurance as FinTech activities. However, Myanmar problems are now well known: capital constraints regulate the retail sector and product constrains regulate the low-quality financial product sector. Access to financial services is low at only 30% having access to formal financial services.

Myanmar faces many challenges, such as a weak regulatory environment, cash-based economy, weak saving culture, low trust in the formal system, inappropriate and insufficient products, capital constraints, limited accessibility, weak institutions, absent infrastructure, and lack of hard infrastructure. However, these challenges can be viewed as opportunities as this is one country that has few legacy systems and historical silo structure that plague other well-developed financial centers.

There have been many improvements in the regulation, with the announcement of the rules for mobile money operators. However, so far, there is no policy regulating the overall payment ecosystem. There are a lot of grey areas, which give space to unregulated activities, such as hundi (unofficial remittance system). Another issue is talent. There is limited local know-how in Myanmar, and it is still difficult to attract foreign talents. Capital is also limited. Lastly, the financial system in itself is still under development: for instance, KYC is still not mandatory for Telcos and Money providers. Myanmar has, since the beginning of 2016, its clearance system, a stock exchange market, and a Payment System (PMU). However, those infrastructures are still under construction.

Regarding expectation of Myanmar's FinTech startups, its financial services industry has begun a significant transformation with the explosive growth of mobile and Internet penetration in a major impact. Mobile operators, partnering with banks, are best placed to expand the reach of financial services in Myanmar to those that are unbanked today, leveraging the significant distribution networks that are already in place.

A nascent financial system is where Western FinTech has been built on top of existing old systems. Burmese FinTech has to start everything from scratch. Equally lacking is a limited talent pool. It was repeated during our interviews: Burmese "FinTech" talent is, so far, a scarce resource. However, some "tech hubs", such as the Yangon MICT Park (and to a lesser extent the Mandalay's one), might well be the future hubs for appealing fresh and innovative spirits.

4.6.1.9 *Cambodia*

In 2016, only 13% of adults have bank accounts with less than 4% with savings, a situation very similar to Myanmar a few years back. However, 100% of the population has mobile phones with 4G coverage with 70% having a National ID. So far, none of the mobile money service providers offer saving products, and none of the MFIs have been able to develop an extensive network of agents to facilitate microdeposits. Wing, Metfone, Banhji, Bongloy, Smart Axiata, and Asia Wei Luy are some notable FinTech companies in Cambodia.

4.6.1.10 *Laos*

Laos has a small and young population of 7 million, but its gross domestic product (GDP) growth is over 7%. Half of its population is under 25 years, with one-third of the population living in urban areas. According to Wear Social's statistics, there is 20% Internet penetration with an unbanked population of close to 60%. Cell phone penetration is 85%. The only FinTech solution is offered by The Banque Pour Le Commerce Exterieur Lao (BCEL) with a secure online payment system for its merchant customers. It has developed a mobile application, BCEL One, to facilitate cardholders' payments and transaction inquiry through mobile or the Internet. The commercial bank, BCEL, and a mobile network operator, UNITEL, are working to start mobile money operations in which people can deposit money to a network of agents of UNITEL who will convert it into e-money and send it to distant locations. Thailand's SuperRich currency exchange platform is attempting to offer a digital platform for foreign exchange, cash transfers, and payment services.

Farming remains the main source of income, and all financial activities are through the informal network of friends and family. Financial literacy, insurance coverage, savings, transactions, and remittances are key to financial inclusion. As for FinTech, the small population is a major disadvantage for Laos.

4.6.1.11 *Brunei*

In February 2017, Authority Monetary Brunei Darussalam (AMBD) had undertaken several initiatives to encourage the growth of FinTech companies and launched AMBD's FinTech Sandbox or a virtual facilitation office. The FinTech Regulatory Sandbox Guidelines were also launched during the event, guidelines which will allow FinTech companies to test their financial products or solutions safely and efficiently while receiving regulatory support from AMBD.

4.6.1.12 *China*

It is worthwhile to highlight that there are two parallel tracks of FinTech development, one in the western economies and another one in the developing economies, such as China and Africa. The FinTech developments in the western economies have emerged, as a result of the GFC, whereas in developing countries such as China, the growth of FinTech is fueled by the appalling state of their financial sector. China is now the world's leader in FinTech with the "Chinese Dragons" overtaking the US in FinTech investment[19] in 2016. Chinese mobile payments were nearly 50 times greater than

[19] "Chinese Dragons" Overtake US in Financial Tech Investment, http://fortune.com/2017/01/23/china-fintech-invest-citi-report/.

those in the US in 2016.[20] China is also a superpower when it comes to online lending, where the country accounts for 75% of the global market. Chinese FinTech companies have also been recognized as the most innovative in 2016, in a ranking of the world's most innovative FinTech companies, four out of the top five slots are occupied by the Chinese companies. The largest Chinese FinTech company, Ant Financial, has been valued at about US$60 billion, making it comparable to UBS, the largest bank in Switzerland.[21]

The Economist[22] has attributed the rapid growth and adoption of such FinTech services in China to

1. **Late-adopter benefits — moving from cash to mobile payments directly:** The middle-class Chinese have always been good online shoppers, ever since the Internet took off. Therefore, they are very comfortable in carrying out financial transactions online and this made them big and early adopters of digital and mobile payments.

   China also had a late-starter advantage. Most developed countries such as US and Singapore have replaced cash for credit and debit cards. China was, until a decade ago, a cash-based economy. The shift to digital and mobile payments was further accelerated by the adoption of smartphones, which allows Chinese consumers to pay for their retail transactions using their phones.
2. **Overlooking small borrowers:** State-owned banks dominate the financial system in China and they have a strong preference for lending to state-owned companies. Furthermore, the lack of a mature credit system in the country for assessing default risks in loans make it difficult for these banks to grant loans and for individuals to obtain loans. This drives consumers to pawn shops or loan sharks which charge exorbitant interest rates. Therefore,

---

[20]China mobile payments dwarf those in US as fintech booms, research shows, https://www.ft.com/content/00585722-ef42-11e6-930f-061b01e23655?mhq5j=e2.

[21]In fintech, China shows the way, http://www.economist.com/news/finance-and-economics/21717393-advanced-technology-backward-banks-and-soaring-wealth-make-china-leader.

[22]Ibid.

the rise of online lending or P2P lending fills the gap for the dearth of loan services to small and medium enterprises (SMEs).

3. **China is a nation of savers:** As a nation of savers, the investment opportunities of the Chinese are limited. They are governed by strict capital controls which limit individuals to no more than US$50,000 each year in foreign currency. Therefore, their savings are either stashed in low-interest-bearing savings accounts or punted on the stock market. There isn't a middle ground.[23] The emergence of FinTech created a new wave of asset managers and money market funds, where savers head online to deposit their money with these FinTech companies. The returns generated by these funds are generally higher than the deposit rate offered by banks without a significant increase in the risks they need to bear. In fact, people's need for such investment opportunities are so great that it took Yu'e Bao only four years, from its inception, to overtake J.P. Morgan's US government money market fund (established since 1993) as the world's biggest funds under management.[24]
4. **Dominant market players:** In China, the tech players, namely the BAT (Baidu, Alibaba, and Tencent), are responsible for making FinTech a huge success in the country. These companies were very successful in their own businesses — such as e-commerce for Alibaba, messaging services for Tencent, and search engine for Baidu and they did not start out by providing financial services to their users. They were popular and had a huge number of users, which allowed them to harness data to capture the spending needs of their users. They were therefore able to provide the financial services that their users needed at the right time. For example, Alibaba allows their shoppers to conveniently borrow small amounts (<10,000 yuan) and Ant Financial, a subsidiary of Alibaba which was responsible for such lending, which was

[23]Risky returns, http://www.economist.com/news/special-report/21697984-china-struggling-unleash-power-stocks-and-bonds-risky-returns.

[24]Chinese money market fund becomes world's biggest, https://www.ft.com/content/28d4e100-2a6d-11e7-bc4b-5528796fe35c?mhq5j=e2.

profitable enough such that it was spun out of Alibaba in 2014.

These market players have been very successful in providing an ecosystem in retaining and attracting new users to their services. For example, users of Tencent's WeChat, have been using the app for texting and subsequently, for doing online and offline payments. The payment functions within WeChat or Ant Financial's Alipay exist elsewhere in the world, but in various apps/platforms: Stripe/PayPal for online shopping, Apple Pay/Android Pay for retail translations, and Venmo for transferring money to social contacts. In China, all these online, offline, and P2P payments are aggregated within the WeChat Pay or Alipay's ecosystem. This explains the rapid proliferation and natural adoption of mobile payments in China.

Many successful FinTech companies in the world such as Ant Financial (Financial Services), Qudian (Qufenqi, Micro Students Loan), Lufax (P2P and Financial Services), Zhong An (Online Insurance), and JD Finance (Supply Chain Financing) are all originally located in China and focus on the previously underserved market segments. Their strategies were not only to take advantage of economies of scale but also to take advantage of economies of scope.

References

Lee, D. K. C. & Deng, R. H. (2018). *Handbook of Blockchain, Digital Finance, and Inclusion*, Vol. 1. Academic Press.

Lee, D. K. C. & Low, L. (2018). *Inclusive FinTech: Blockchain, Cryptocurrency and Initial Crypto-Token Offering*. World Scientific Publishing.

Chapter 5

Platform Strategy and Sharing Economy

"The economy of today and tomorrow will rely on a platform and ecosystem approach."

—Jack Ma

5.1 Introduction

In today's networked era, the ability to create a successful platform determines the competition landscape of many industries. With an aggregate market value of US$4.3 trillion (Evans and Gawer, 2016), platform businesses constitute some of today's most highly valued companies in a growing number of industries, such as urban transportation (Uber, Lyft), hospitality (Airbnb), human resource management (TaskRabbit, Freelancer), social networking (Facebook, WeChat), retail (Amazon, eBay, Alibaba), mobile payment (Apple Pay, WeChat Pay), and clean energy (Sungevity, SolarCity, EnerNOC). Predominant in the US and China, these platform companies have an increasing global presence. By constructing a digital platform, other individuals and/or businesses can simply connect their business with the platform, build products and/or services on it, and realize value co-creation. In this way, value is created by users, for other users, and the platform merely serves as an interface upon which this value creation and consumption can take place at a scale.

This chapter introduces the rise of the platforms and addresses some of the key issues concerning the impact, technology, and development in platform markets. As one of the fastest growing type of platforms, the sharing economy platform has gained increasing research attention in recent years. This chapter also examines characteristics on these on-demand sharing platforms and the opportunities and challenges regarding their effective application.

5.2 An "Age of Platforms"

In an "age of platforms", the trend is accelerating rapidly across multiple industries. The connectivity and the ability to "plug-and-play" is a defining characteristic of Platform Thinking. For instance, the success stories of Apple and Android's ecosystems have been built on the app store on which external developers create value. Besides the technological sector, retailers are also shifting their focus from being a part of traditional distribution channels to engagement platforms. The use of platform thinking extends beyond the online retailers like eBay, Etsy, and Amazon. Traditional retailers like Walmart have also extended their competition to the digital space, teaming up with Google.

In light of the recent growth of platforms in size and scale, they are also contributing in significant ways to the economies. In 2017, the top 50 platform companies have a combined market capitalization of close to US$5 trillion. The "super-major platforms", such as Apple, Google, Facebook, and Microsoft, have a combined market capitalization of US$3.5 trillion (The MIT Initiative on the Digital Economy (IDE), 2017). China's "super-major platforms", such as Alibaba and Tencent, are also leading the wave. A study from Boston University and Massachusetts Institute of Technology (MIT) shows that companies that open their application program interface (API) experience a 4% increase in market capitalization on an average (IDE, 2017). Likewise, companies that make developer portals available tend to have lower research and development costs along with improved revenue.

Platforms have progressed from being an innovative business model to a disruptive force across multiple industries, leading to

Table 5.1: Platform Ecosystem

	Consumers Behavioral changes among consumers are driving the demand for platforms and "X-as-a-service"				
Customer Interface	**Service Platforms** Connect and facilitate transactions between individuals and/or businesses				
Industry clusters	**Transportation** For example, Uber	**Hotel and accommodation services** For example, Airbnb	**Retail** For example, Amazon	**Entertainment** For example, Spotify	...
Enabling platforms	**Payment,** **Distribution,** ...				

Source: Lacy *et al.* (2016).

platform-based ecosystems driven by customer demand. Businesses across industry clusters are entering existing platforms as enablers or are collaborating to build new products, services, and customer experiences. Table 5.1 summarizes the various components of the platform ecosystem that are changing the traditional rules of market competitiveness.

In the platform ecosystem, previously "independent" products and service suppliers are now facing blurring industry boundaries and converging markets. Former competitors are increasingly working closely together, and former collaborators could evolve into competitors. The traditional vertical industry clusters are increasingly competing with offerings of horizontal platforms (e.g., Spotify on iOS vs. iTunes and Netflix on Prime vs. Amazon). The platform economy leads to both expanded competitive circle and bigger addressable market. With this heightened level of industry convergence, companies must find out meaningful ways to reposition in the new competitive space.

5.3 Technologies Behind the Rise of Platforms

The rise of platforms has been driven by three transformative technologies: cloud, social networks, and mobile devices. The result is a globally accessible network of entrepreneurs, businesses, talents, and

Table 5.2: Drivers Behind the Rise of Platforms

Drivers	Impacts
Cloud	Enables a global infrastructure for production, allowing anyone to create content and applications for a global audience.
Social networks	Connects people globally and maintains their identity online.
Mobile devices and IoT	Connects users to the global infrastructure anytime, anywhere.

Source: Bonchek and Choudary (2013).

customers who are readily available to build technologies, develop business applications, contribute contents, and consume goods and services (see Table 5.2).

5.4 Common Types of Platforms

According to Deloitte Insights, three major categories of platforms are becoming increasingly prominent: aggregation platforms, social media platforms, and mobilization platforms as presented in Table 5.3 (Hagel, 2015).

5.4.1 *Aggregation Platforms*

Aggregation platforms bring together a wide range of relevant resources and connect users of the platform with the most appropriate resources. These platforms tend to be very transactional/task-focused and tend to operate on a hub-and-spoke network. All transactions are mediated by the platform owner.

5.4.1.1 *Subcategories of aggregation platforms*

- Data/information aggregation platforms, e.g., scientific databases.
- Marketplace platforms, e.g., eBay, Airbnb, App Store etc.
- Contest platforms — where one can post a problem/challenge and offer a reward to the participant who provides the best solution.

Table 5.3: Three Common Types of Platforms

Common Types of Platforms	Platform Characteristics
Aggregation platforms	• Connect users of the platform with the most appropriate/relevant resources • Transaction- or task-focused • Hub-and-spoke model
Social media platforms	• Build/reinforce long-term relationships across participants on the platform • Mesh networks
Mobilization platforms	• Move participants to act together to accomplish something beyond the capabilities of any individual participant • Build long-term relationships

Source: Hagel (2015).

5.4.2 *Social Media Platforms*

Social media platforms differ from aggregation platforms on some key dimensions. First, they build and reinforce long-term relationships across participants with a common interest on the platform, instead of a transaction/task. Second, they tend to foster mesh networks of relationships — participants connect with each other over time in various ways that usually do not involve the platform owner. Facebook and Twitter are examples of popular social media platforms.

5.4.3 *Mobilization Platforms*

A key focus of mobilization platforms is to connect with, and mobilize, a given set of participants and resources to achieve a shared goal, taking common interests to the level of execution. The most common form of mobilization platforms is "process network" platforms that bring together participants in extended business processes like supply chains or distribution operations that help to select and direct participants who need to collaborate with each other. For example, popular examples of mobilization platforms include the global sourcing company Li & Fung and open-source software platforms like Linux.

5.5 Factors That Determine the Success of a Platform

The success of a platform strategy is determined by three factors: connectivity, social gravity, and matchmaking capability (Bonchek and Choudary, 2013). Refer to Table 5.4 for a summary.

5.5.1 *Connectivity*

Platforms need to create an infrastructure that makes it easy for participants to plug into the platform and enable interactions among participants. For example, Apple provides app developers with the OS and underlying code libraries; Taobao provides online stores to its shop owners.

5.5.2 *Social Gravity*

Platforms need to attract participants. Both producers and consumers must be present to achieve critical mass. For instance, Apple needs to attract both developers and users. Similarly, eBay needs

Table 5.4: Factors That Determine the Success of a Platform

Factors That Determine the Success of a Platform		Examples
Connectivity	How easily various parties can "plug-and-play" on the platform in order to generate and share contents, make transactions, and/or interact with other participants	Wikipedia provides contributors with the tools to collaborate on an article
Social gravity	How well the platform attracts producers/suppliers and consumers	eBay needs to attract both buyers and sellers to achieve critical mass
"The Matchmaker"	How well a platform fosters trust, matches the demand with supply, and facilitates the exchange and co-creation of value	Uber matches drivers with the transportation needs of passengers

Source: Bonchek and Choudary (2013).

both buyers and sellers. Platform builders must pay attention to the design of incentives, reputation systems, and pricing models to motivate participation of both parties. They can leverage social media to harness the network effect for rapid growth.

5.5.3 *Matchmaking Capability*

A matchmaker fosters the flow of value by connecting producers/suppliers and consumers. Data analytics are essential for matchmaking and distinguish platforms from other business models. The matchmaker captures data about the participants and optimizes connections between producers and consumers. For example, Google matches the supply and demand of online content, while marketplaces such as eBay match buyers to sellers of relevant products.

Not all platforms place the same emphasis on connectivity, social gravity, and matchmaking capability. For example, eBay and Airbnb have focused more on the social gravity and matchmaker role. Facebook has focused on connectivity and social gravity. A network effect (also called demand-side economies of scale) is the positive effect that an additional user of a product or service has on the value of that product/service to others. This characteristic is a major advantage of platform businesses. When a network effect is present, the value of a product/service increases according to the number of users using it. The platform ecosystem realizes the interconnection between two or more parties, thus forming a strong network effect and enhancing the overall value with the participation of more members. In the case of Uber, more drivers on the platform can translate into shorter wait times and wider geographical coverage for passengers within a city. Strengthening one's network and solidifying customer stickiness is thus the key to business success.

5.6 The Rise of the Sharing Economy

One of the most rapidly growing technology-mediated business models is the sharing economy. The sharing economy has emerged in recent years as a disruptive approach to traditional business-to-business and business-to-consumer business models. PricewaterhouseCoopers (2015) estimates that global revenues from sharing in just five major sectors — travel, car sharing, finance, staffing, music, and video streaming — will increase from US$15 billion in 2015 to US$335 billion by 2025.

The sharing economy is an economic and social system that enables shared access to underutilized assets from skills/services to products for monetary or non-monetary benefits. The assets targeted by these startups range from spare time for everyday tasks (TaskRabbit) and spare time and automobiles to drive people around (Uber, Lyft) to extra rooms (Airbnb) and occasionally used tools and other household items (Streetbank, SnapGoods/Simplist).

Peer-to-peer networks can grow exponentially through the power of platform dynamics and network effects. Sharing economy startups also reflect the broader "servitization" trend. Instead of selling products outright, companies can expand their potential markets by renting access to products that people used to purchase. For instance, automakers such as Daimler and BMW offer transportation as a service instead of selling automobiles (Cusumano, 2015).

The sharing economy system consists of suppliers, consumers, and the platform that is connecting them (see Figure 5.1). Under such a system, the practices of sharing, exchange, or rental of goods and services to others do not involve the transfer of ownership.

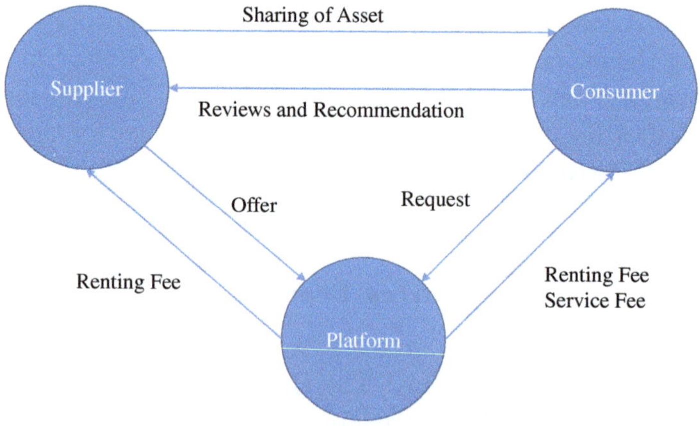

Figure 5.1: The Sharing Economy System

The sharing economy has a transformative effect on how goods and services are being provided and consumed. The main advantage of the sharing economy is the improved efficiency and effectiveness by reducing transaction costs and increasing the rate of utilization of goods and services. For instance, private automobiles usually remain idle more than 90% of the time.

Carpooling applications have provided an excellent way of large-scale sharing of both the cost of commuting and leveraging an expensive asset. Besides cost savings and convenience, it has enabled environmental benefits on a large scale. By allowing access to underutilized resources at a fraction of the cost to someone who cannot or does not want to buy new products and the chance of making an extra income for those who already own such underutilized resources, the sharing economy is seen as instrumental in addressing issues such as overconsumption and income inequality.

5.7 Drivers of the Sharing Economy

PricewaterhouseCoopers (PwC)'s projections in 2015 show that five key sharing sectors — travel, car sharing, finance, staffing, and music and video streaming — have the potential to increase global revenues from roughly US$15 billion in 2015 to around US$335 billion by 2025. There are many drivers behind the success of sharing economy platforms (see Table 5.5).

5.8 Categorization of the Sharing Economy

Recently, sharing-based businesses have evolved from simple peer-to-peer lending initiatives to complex platforms and networks of people and companies interacting for the collective use of extant or new resources — asset hubs (see Figure 5.2) (Muñoz and Cohen, 2017).

Muñoz and Cohen (2017) developed a set of dimensions from which the sharing economy business models could be evaluated and differentiated: (1) platforms for collaboration, (2) underutilized resources, (3) peer-to-peer interactions, (4) collaborative governance, and (5) technology reliance. Table 5.6 elaborates on each of these dimensions while Table 5.7 provides an analysis on major market

Table 5.5: Drivers of the Sharing Economy

Drivers of the Sharing Economy	Characteristics
Excess capacity — Access over ownership Excess capacity makes the sharing economy possible.	• Cars, just like other "lumpy" goods — apartments, office spaces, bikes, party equipment, computers, clothes, books etc. — have to be purchased in units that exceed the buyer's immediate consumption needs, which bring with them a good deal of excess capacity — lumpiness. • This also applies to physical and intellectual labor. An experienced worker's ability goes unused much of the time.
IoT and location-based mobile technologies Location-based mobile devices allow both consumers and suppliers to quickly make and respond to requests for goods and services.	• A precise, real-time measurement of spare capacity and the ability to dynamically connect that capacity with those who need it. • What made Uber, Grab, Airbnb, and all the other sharing economy companies possible is the combination of big data analytics, low-cost cloud storage, prevalence of social media, and widespread use of location-based mobile devices.
Light on infrastructure investment	• Conventional rental-based business models demand not just excess capacity at a scale but also heavy infrastructure. Hotels have to maintain properties and provide other services. Similarly, car rental companies have to maintain and store cars that are not in use. Sharing economy platforms like Airbnb, on the other hand, are asset-light.
Online reviews — An economy built on trust Platforms facilitate this through user-friendly review and rating systems.	• Most sharing companies establish trust through online reviews — one form of crowdsourcing. Before a consumer uses an Uber driver or rents an Airbnb room, he/she typically checks out what others who've used the particular service have to say.

(*Continued*)

Table 5.5: (*Continued*)

Drivers of the Sharing Economy	Characteristics
Others	• The sociability that emerges via peer-to-peer collaborative consumption. • Difficult economic times, financial crises, and recessions have contributed to increased frugality; both suppliers and consumers can benefit from the sharing economy system. • Rise of urbanization and limited urban space, population growth, and urbanization have increased sensibility about sustainability, leading to an increased need for alternative modes of consumption.

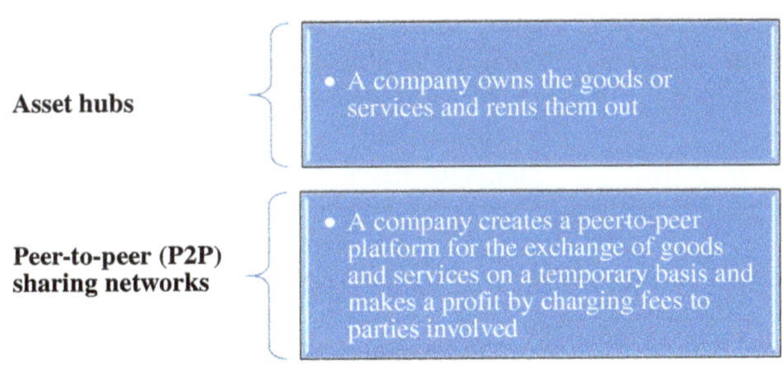

Figure 5.2: Categorization of the Sharing Economy

Source: Taeihagh (2017).

players on the three most prominent characteristics of the sharing economy: peer-to-peer, underutilized resources, and access over ownership.

5.9 Case Studies

5.9.1 *Uber*

"Uber" comes from "über" in Germen, means "super". Being a technology company, Uber allows consumers to request car

Table 5.6: Dimensions of the Sharing Economy Business Model

Dimensions	Business Model
Platforms for collaboration (Chase, 2015; Cohen and Muñoz, 2016)	Platform as a defining feature of sharing economy initiatives.
Underutilized resources (Chase, 2015; Cohen and Kietzmann, 2014)	Peer-to-peer sharing — encourages peers to optimize the use of their own personal (previously purchased and already existing in circulation) items (e.g., vehicles). Business-to-crowd model — relies on the purchase and ownership of new items (e.g., vehicles) by the intermediary, placed throughout a city with the intent of encouraging the optimization of these new items.
Peer-to-peer interactions (Chase, 2015)	P2P interactions as a critical element of sharing business models.
Collaborative governance (McLaren and Agyeman, 2015)	The importance of collaborative and participatory governance models for community-based models within the sharing paradigm.
Technology reliance (Belk, 2014; Sundararajan, 2016)	Web technologies — a primary connecting point. Location-based, geo-referenced data — to facilitate more efficient transactions.

Table 5.7: Online Platforms Associated with the Sharing Economy

Product/Service Category	Examples	Traditional Channel	Peer-To-Peer	Underutilized Resources	Access over Ownership
Ride sharing	Uber, Lyft	Taxis	√	√	√
Accommodations	Airbnb	Hotel	√	√	√
Personal services	Taskrabbit	Classified ads, Temp agency	√	√	√
Car/bike sharing	Zipcar, Ofo	Rental car/bike	×	×	√
Crowd funding	Kickstarter, Indiegogo	Venture Capital	√	×	×
Money lending	Funding Circle	Banks	√	×	×

Source: Gulker (2017).

transportation or food delivery via the Uber mobile app or the Uber mobile website. Uber drivers use their own personal cars, although drivers can rent a car to drive with Uber. Uber keeps 20% of the booking revenue of every ride. Its service is available in over 528 cities in close to 70 countries worldwide and growing. It was the 48th-most powerful company in America in 2014 with its rapidly growing revenues. Its US$68 billion valuation in 2016 makes it the world's most valuable private company.

Uber holds a vast database of drivers in all of the cities it covers, so when a passenger asks for a ride, they can instantly match him/her with the most suitable drivers. Fares are calculated automatically, using Global Positioning System (GPS), street data, and the company's own algorithms which make adjustments based on the time that the journey is likely to take. This is a crucial difference from regular taxi services, where customers are charged for the time the journey takes, not the distance covered.

5.9.2 *Surge Pricing*

This is an implementation of "dynamic pricing", similar to that used by hotel chains and airlines to adjust price to meet demand, although rather than simply increasing prices at weekends or during public holidays, it uses predictive modeling to estimate demand in real time. These algorithms monitor traffic conditions and journey times in real time, meaning prices can be adjusted as demand for rides changes and traffic conditions mean journeys are likely to take longer. This encourages more drivers to get behind the wheel when they are needed and stay at home when the demand is low.

Compared to conventional taxis, four factors are likely to contribute to the high-capacity utilization rate of Uber: (1) Uber's more efficient driver–passenger matching technology, (2) the larger scale of Uber than taxi companies, (3) inefficient taxi regulations, and (4) Uber's flexible labor supply model and surge pricing more closely match supply with demand throughout the day.

Users can rate drivers, and vice versa, through its inbuilt rating system — to build up trust and allow both parties to make informed decisions about who they want to share a car with.

5.9.3 *Airbnb*

Airbnb was founded in 2008, based in San Francisco. It is a leading community-driven hospitality company where people can list, discover, and book unique accommodations around the world. As an online platform matching room seekers to homeowners, it connects more than 1,500,000 listings in 34,000 cities/190 countries. The guest service fee for homes ranges between 0% and 20% of the booking subtotal and is calculated using a variety of factors — these include the reservation subtotal, the length of the stay, and characteristics of the listing. In general, the service fee gets lower as the reservation cost gets higher. The host service fee for homes is generally 3%.

With its highest market valuation and number of room listings in the hotel industry, Airbnb is becoming a real threat to the conventional hotel businesses. Its revenue growth is also outpacing that of other hotel-booking sites such as booking.com, Expedia, hotels.com, etc.

The Growth Story:

- Solve a Real Problem — Helps travel enthusiasts find a comfortable and affordable accommodation; it also helps those who have additional space to earn a passive income by renting it out;
- Put User Needs First — Advance payments feature, professional and attractive photos, more personal, better descriptions;
- Highly differentiated listings;
- Airbnb Social Connections.

By comparison, the world's largest hotel chains have fewer rooms, much slower growth rates, and much lower valuations compared to their revenues. For hotel chains to add rooms, they can either acquire an existing hotel or (such as Hilton's US$13 billion acquisition of Starwood) set up new hotels themselves. Therefore, expansion is financially risky for the investors. By contrast, Airbnb grows its portfolio mostly on the power of network effects. As the company invests more on advertising (for example, Airbnb's ads on Expedia.com), more people offer their rooms and more people use the service through network effects.

5.10 Competitive Strategies for Traditional Businesses

How do conventional businesses compete with these platforms? First, conventional businesses should make sure that the playing field is level (Cusumano, 2015). Traditional businesses should highlight the potential risks behind the sharing economy business model to the local government officials and make sure that these platforms are not violating the rules for licenses for subletting, taxes, insurance, training of the service providers, safety features, and other relevant regulations. Second, traditional businesses can compete by differentiation through services and unique value propositions. Traditional hotels can compete by providing a more standardized, consistent, and reliable level of service that Airbnb cannot match. For example, Airbnb lags in confirming reservations, making it difficult to make definite travel plans. Hotels can target the business travelers and corporate/personal events, such as trade conferences or weddings. They can link customers with a broad variety of vendors for tour packages, transportation, and other services, tailored to the locality and the occasion. For transportation business, taxis are highly regulated and restricted, though they are not always available, especially in rural communities or during peak hours. Like hotel chains, taxi companies can offer standardized service, safety, and fixed price that Uber will not always be able to match. Uber drivers may not be familiar with the roads. They can also decline a booking of service. Its "dynamic pricing" has been widely criticized. On the other hand, taxis are obligated to offer standardized prices and service to any passenger. Yet, the challenge for differentiation is tougher in the taxi market, since a car ride does not require the same level of discretion as choosing a hotel. Lastly, traditional companies can also adopt platform thinking. For example, ComfortDelGro has introduced a mobile app that enables potential customers to order taxis and pay via credit card or digital payment system.

While sharing economy platforms like Airbnb and Uber get all the media attention due to their unprecedented valuations and market penetration, they also have drawn some tensions and criticism from a wide range of stakeholders who claim these models unfairly compete in an unregulated environment, fail to meet minimum quality and safety standards, exploit "on-demand" workers, and, in the case of some listings with Airbnb, have detrimental impacts on local neighborhoods and quality of life in cities (McLaren and Agyeman, 2015). For instance, actors in the hospitality and transport sectors are lobbying local governments in places like New York City and Berlin to stop the growth of Airbnb and Uber, respectively.

5.11 Conclusion

In the face of the digital revolution, the contemporary businesses are creating new ways of provisioning goods and services and opportunities for peer-to-peer technology-mediated exchanges. Platforms help connect resources/producers and participants on an as-needed basis and scale participation and collaboration. Companies need to reassess customer needs in the age of digital disruption and address them in the most productive way. Businesses need to reposition themselves as a critical part of the integrated solutions and value chain. Well-designed business models and platforms can create and capture new economic value and scale the potential across entire ecosystems.

The unprecedented growth rates for sharing platforms have clearly raised the attention of many established industry incumbents. Some startups have already run into legal and regulatory hurdles from city governments, courts, and traditional unions or lobbies that want to restrict them or shut them down. Yet, not all traditional industries view the sharing economy exclusively as a threat. Large car manufacturers, such as BMW, Daimler, Volvo, and Ford, have decided to test the car-sharing space. Most notably in the case of BMW's Drive Now, the company expects that the increased exposure to their vehicles to a new segment not normally targeted by the luxury brand may result in future sales to current car-sharing customers (Muñoz and Cohen, 2017).

References

Belk, R. (2014). You Are What You Can Access: Sharing and Collaborative Consumption Online. *Journal of Business Research*, 67(8), 1595–1600.

Bonchek, M. & Choudary, S. P. (2013). Three Elements of a Successful Platform Strategy. *Harvard Business Review.* Retrieved from https://hbr.org/2013/01/three-elements-of-a-successful-platform.

Chase, R. (2015). Peers Inc: How People and Platforms Are Inventing the Collaborative Economy and Reinventing Capitalism. *Public Affairs*, New York.

Cohen, B. & Kietzmann, J. (2014). Ride on! Mobility Business Models for the Sharing Economy. *Organization & Environment*, 27(3), 279–296.

Cohen, B. & Muñoz, P. (2016). Sharing Cities and Sustainable Consumption and Production: Towards an Integrated Framework. *Journal of Cleaner Production*, 134 (A), 87–97.

Cusumano, M. A. (2015). How Traditional Firms Must Compete in the Sharing Economy. *Communications of the ACM*, 58(1), 32–34.

Evans, P. C. & Gawer, A. (2016). *The Rise of the Platform Enterprise: A Global Survey.* The Emerging Platform Economy Series, The Center for Global Enterprise.

Gulker, M. (2017) *The Real Innovation of the Sharing Economy.* Retrieved from https://www.aier.org/article/real-innovation-sharing-economy.

Hagel, J. (2015). *The Power of Platforms.* Deloitte Insights. Retrieved from https://www2.deloitte.com/insights/us/en/focus/business-trends/2015/platform-strategy-new-level-business-trends.html.

Lacy, P., Hagenmueller, M., & Ising, J. (2016). *Platform Strategies: How the Rules of Competitiveness Have Changed in the Era of Ecosystems.* Accenture Strategy, Retrieved from https://www.accenture.com/t20160916T010708Z__w__/sg-en/_acnmedia/PDF-27/Accenture-Strategy-Platform-Strategies.pdf.

McLaren, D. & Agyeman, J. (2015). *Sharing Cities: A Case for Truly Smart and Sustainable Cities.* MIT Press.

Muñoz, P. & Cohen, B. (2017). Mapping Out the Sharing Economy: A Configurational Approach to Sharing Business Modelling, *Technological Forecasting & Social Change*, 125, 21–37.

PWC (2015). The Sharing Economy, Consumer Intelligence Series. Retrieved from https://www.pwc.com/us/en/technology/publications/assets/pwc-consumer-intelligence-series-the-sharing-economy.pdf.

Sundararajan, A. (2016) *The Sharing Economy.* MIT Press.

Taeihagh, A. (2017). Crowdsourcing, Sharing Economies and Development. *Journal of Developing Societies*, 33(2), 191–222.

The MIT Initiative on the Digital Economy (IDE) (2017). *2017 MIT Platform Summit Report.* Retrieved from http://ide.mit.edu/sites/default/files/2017-MIT-Platform-Summit-Report.pdf.

World Economic Forum (2017). *Digital Transformation Initiative: Unlocking B2B Platform Value.* Retrieved from http://reports.weforum.org/digital-transformation/wp-content/blogs.dir/94/mp/files/pages/files/wef-platform-report-final-3-26-17.pdf.

Chapter 6

Digital Transformation and the "New Marketing Normal"

6.1 Introduction

New technologies with its capability to exchange information from physical objects are blurring the lines between the physical and digital world. Over the past decade, the marketing landscape is undergoing radical changes due to the proliferation of digital technology and customer mindset and value shifts. Marketers need a holistic approach to reach customers, acknowledging the importance of relevant and authentic conversations with customers and their growing power in the digital era. Ira Kaufman and Chris Horton published a book in 2015 about the leading role of the marketing function as the principal agent of change in the digital transformation throughout the organization.

Because the basic premise of how the customers would interact with companies and their products, services, and brand is changing, marketers need to adopt a new approach and framework. Digital transformation, that is impacting the entire organization, starts with a number of fundamental changes to the existing marketing function, such as deeper and more interactive online customer engagement, more relevant conversations with customers, and more authentic brand experiences. We refer to this as the "New Marketing Normal" (Kaufman and Horton, 2015) (see Table 6.1).

Table 6.1: Foundation of the "New Marketing Normal"

Marketing Functions	Key Change Elements
Segmentation	From target consumers to building consumer networks
Branding	From strategic branding to social brands
Product Features	From product experience to customer experience (CX)
Sales Process	From sales funnel to digital involvement cycle
Customer Decision Process	From consumer behavior to influence marketing

Source: Kaufman and Horton (2015).

Table 6.2: Key Questions for Success in the "New Normal"

Key Questions	Examples
How embedded is digital solution in the business	Amazon's smart home automation systems
"Micromoments" that matter	Google searches for "[something] near me"
Data-driven marketing	Facebook ads targeting features allow marketers to reach people based on their demographics, location, interests, and behaviors
Integrated channel solutions	Online ordering with a ship to the store for pickup option by Home Depot and Walmart
"Faster, more agile, and thinking 10×"	FXMirror, an augmented reality mirror lets brides-to-be see how they look in gowns without stress and hassle

Source: Soat (2016).

"To be a great marketer, you must be a great digital marketer, says Jim Lecinski, VP of US sales and service at Google. To be a great digital marketer, you must be a great mobile marketer" (Soat, 2016). According to Jim Lecinski, there are five key questions marketers should be asking now to winning in the new normal (see Table 6.2): First, how embedded are digital elements in your product offering and in your business? Amazon's smart home systems that control room temperature and security and Hilton's digital room keys are all based on mobile devices. Second, how is your brand present in the micromoments and the moments that matter, especially on mobile? More and more consumers conduct research about a product on the go or even in the store. A micromoment is a small opportunity to connect with a consumer at the precipice of purchase by providing localized and relevant information, such as Google searches for "[something] near me". Third, is your company evolving the approach to take advantage of the data to make better decisions and achieve better performance? With artificial intelligence, cloud computing, and machine learning, the analytic capabilities have provided new ways to understand consumer behaviors and target them adaptively. Fourth, is your company breaking down channel barriers and integrating different marketing channels to create a holistic customer experience? Brands like Sephora have started to embed purchase options inside YouTube videos. Alibaba's Singles' Day Shopping Festival is a combination of commerce and entertainment. Alibaba offers augmented reality mobile gamification: The Tmall "Catch the Cat" game will allow consumers to earn special promotion coupons and prizes every time they catch the "Tmall mascot" via Alibaba's Taobao and Tmall apps. Consumers can collect "red packets" through the Tmall app and Tmall virtual stores, as well as share these red envelopes with friends (Coleman, 2017). Lastly, are you able to foster what we call being "faster, more agile, and thinking 10×?" According to Lecinski, many marketers get stuck in a "10% mindset" — a small, achievable, incremental growth, 10% at a time. They must instead think 10×, striving to create a radical, possibly overly ambitious solution. Because of the new way that consumers gather information about brands, the marketers' job becomes informing them about the company's brand and helping them make greater choices. Drawing upon the understanding of these five key questions in the digital age, marketers need to rebuild the existing foundation to reflect these changes. In the New Marketing Normal, marketers can leverage on various digital tools and platforms to engage in timely and actionable data-driven targeting.

6.2 A History of Digital Marketing

The advent of digital marketing can be traced back to the days of the 1980s, when the Internet as a set of network technologies emerged as a way to interconnect across academic and military computer networks.

In 1981, the world's first personal computer was launched by IBM. In 1989, the storage capacity of the computers was only 100mb. It did not become widespread until it was commercialized in the early 1990s.

The 1980s was also the time when companies recognized the importance of managing customer relationship rather than just implementing the operational aspects of pushing products. Some marketers started to include list brokering and database marketing. This was the year when it started to become a practice of maintaining a database of prospects, customers, and commercial contracts. By 1986, ACT!, a contact and customer management company, introduced the first database-marketing software to the business world. This software allows the storage and management of a huge volume of customer information. Robert Kestenbaum and Robert Shaw (known as the father of marketing automation) developed multiple landmark database-marketing solutions for big national businesses, such as Barclays. These database-marketing solutions contained many contemporary features, including campaign management, sales channel automation, contact strategy optimization, marketing resource management, and marketing analytics. Though many stages of the process were manual, the emergence of digital customer databases has changed the dynamics of the buyer–seller relationship.

The modern Customer Relationship Management (CRM) software came in to play in the 1990s. These CRM tools tracked the interactions between sales force and customers and provided companies the important information regarding inventory control. CRM was massively upgraded in the late 1990s when vendors such as SAP and Oracle came into the market.

Display advertising is a form of graphical ad embedded into a webpage, typically including a combination of static/animated images, text and/or video designed to convey a marketing message and/or cause the user to take an action. Banner advertising, also known as clickable display ads, took off between 1994 and 1997. Advertisers were charged an upfront cost to occupy the digital real estate for a set time period, which was very different from today's pay-per-click (PPC) model. Users enticed to click the banner were transported to a landing page from the advertisers. In 1995, WebConnect, an ad agency that specialized in online ads, began helping their clients identify websites that their ideal consumers visited. This led to the beginning of targeted ad placement where their target demographics were more likely to see them. They also capped the number of times a particular user was shown a single banner ad to prevent "banner fatigue". Ad frequency capping is still a common display that ad tactic advertisers use today. In 1996, DoubleClick emerged as one of the first ROI tools for banner ad campaigns. They offered a new service called dynamic advertising reporting and targeting (DART), which enabled advertisers to track how many times an ad was viewed and clicked across multiple websites. Advertisers could track how their ads were performing, analyze the results, and optimize a live campaign. For example, if a marketer noticed their ad was underperforming on one particular website, they could remove the ad and devote those resources to another website where the ad was performing better. DoubleClick also pioneered a new pricing model for online advertising: cost per impression (CPM). Instead of paying a flat fee to host banner ads for a predetermined time period, with improved ad tracking, banner pricing transitioned toward an ROI-based charging model.

In 1998, Google, currently the most popular search engine, came into the market. It plays the most significant role in optimizing digital marketing campaigns and search marketing. Google patented a highly responsive and effective search engine using a relevancy algorithm called PageRank. In the same year, another search company called Goto.com (which was later acquired by Yahoo) came into the scene with a brand new marketing approach — keyword-based search engine marketing. Search engine marketing refers to sponsored links appearing in search engine results pages (SERPs) when users type in a particular keyword that the marketer has purchased. Advertisers bid for specific keywords as a way to reach their users. When advertisers won the bid for certain keywords, their sponsored links would appear with the SERPs when users enter those keywords. Goto.com pioneered the pay-for-placement model, which eventually evolved into pay-per-click business model, whereby advertisers only paid for the ads when users click on their sponsored links. However, the user experience of paid search suffered, when it was difficult for users to differentiate paid from organic content. In 2000, Google introduced AdWords, creating a sponsored search experience that generated revenue without compromising the quality and relevancy of search results. Today, using AdWords, marketers can display sponsored links above or beside the organic search results on the SERP. The mid-2000s saw a change in the behavior of the customers as people began to research their product online using search engines before the actual purchase.

Modern social networking platforms started to gain popularity in the early 2000s. Friendster (2002), then MySpace (2003), Hi5 (2003), LinkedIn (2003), Facebook (2004), and YouTube (2005) represented the early wave of social networking sites, creating an entirely new way for people to communicate and share contents with each other across geographical boundaries. Today, there is a tremendous variety of social networking platforms, and many of them allow cross-posting. Facebook and Twitter both became the most prominent social media among users throughout the world. Other sites like Tumblr, Foursquare, and Pinterest began to fill specific social networking niches.

6.3 The Converged Media Imperative

Marketers have adopted a three-category typology for online media activities: paid, owned, and earned media (Corcoran, 2009; Goodall, 2009; Stephen and Galak, 2012). Paid media refers to media activity that a company (or its agents) generates. Owned media refers to media activities that a company (or its agents) generates in channels it controls. Common forms of owned media include press releases, brochures, and posts made by company representatives on an official company blog or website. "Earned media" refers to media publicity that a company does not directly generate, such as press mentions in traditional media and online community posts in consumer-generated social media. Click-through rate (CTR) is the ratio of users who click on a specific link to the total number of users who view a page, email, or advertisement. It is commonly used to measure the success of an online advertising campaign for a particular website, as well as the effectiveness of email campaigns (see Table 6.3). Organic (unpaid) CTR is usually above 65% on desktop and close to 40% on mobile. Paid CTR is about 3–4% on both platforms (SparkToro, 2018). Figure 6.1 provides a graphical presentation of the three types of media.

The popularity of social media and the increased emphasis marketers place on social approaches have made it more important for firms to understand how multi-channel-earned media affects their sales (Stephen and Galak, 2012). Relying too heavily on one media stream can leave the company vulnerable should an unexpected change reduce its marketing output. Thus, it is important to get a balance of sources and consider all three types of media when developing marketing communications strategies. Figure 6.2 shows the marketing budget split by paid, owned, and earned media.

The emergence of new media extends the marketer's role well beyond the allocation of budgets and channels. Marketers today require a deep understanding of how consumers engage with different types of media at each stage of the customer journey toward a purchase decision. What's more, these different kinds of media are related and interact with one another, so marketing plans and capabilities must adapt and evolve. According to Edelman and Salsberg (2010), paid, owned, and earned media are evolving in three primary ways. First, different types of media are becoming more integrated. Paid media will increasingly serve as feeders into owned media hubs, where marketers can offer a more engaging experience, generate consumer interests, and pivot into an ongoing and more targeted stream of interactions with users or prospects. Companies could push out special deals via social media platforms or company mobile application at a minimal variable cost, reducing its reliance on expensive paid media while fostering closer relationships with consumers. Second, new publishing models are emerging because of the increasing complexity of consumer needs. Marketers are partnering with media publishers to create unique contents that build deeper connection. Branded content, sometimes also known as branded

Table 6.3: Elements of Digital Marketing

Type	Elements of Digital Marketing	Pros	Cons
Paid Media	• Display/banner ads • Paid Search Ads (e.g., Google AdWords) • Social network advertising (e.g., Facebook ads) • Electronic direct mail (e.g., email advertisements)	• On Demand • Immediacy • Highly scalable • High degree of control • Measurable	• Clutters the user experience • Declining response rates • Poor credibility compared to organic/earned media
Owned Media	• Company/brand website • Company/brand blog • E-books • Company-owned social media pages/accounts (e.g., Twitter account, Facebook brand page)	• Control • Cost efficiency • Longevity • Versatility • Niche audiences	• No guarantee of results • Company communication not always trusted • Takes time to generate scale
Earned Media	• Organic search placement • Social media posts (e.g., status updates, tweets) • Online ratings and reviews (e.g., Yelp.com for restaurants, Amazon.com for products) • Social mentions • Online WOM referrals	• Most credible • Plays a key role in most sales • Transparent and sustaining	• No control • Can be negative • Unpredictable scale • Hard to measure

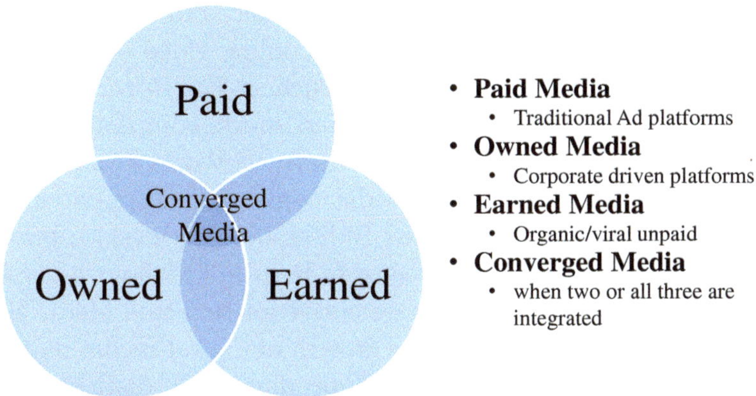

Figure 6.1: Types of Media

Source: Kaufman & Horton (2015)

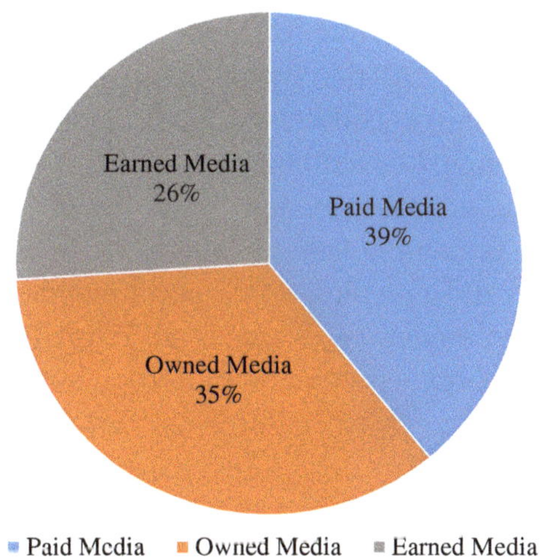

Figure 6.2: Digital Marketing Budget Split by Paid Owned and Earned Media

Source: Oracle Marketing Cloud (2015)

entertainment, refers to media content that is sponsored or produced by a marketer. The energy drink market leader Red Bull has relied heavily on branded content as part of its marketing strategies. The company operates several Media House studios, which coordinate the production and distribution of original content targeted toward the interests of young adults, such as music and extreme sports. Alongside digital media content such as online video (via platforms such as Red Bull TV) and print media such as *The Red Bulletin*, Red Bull has also organized music events and sports competitions which carry its name, like the Red Bull Air Race World Championship and Red Bull BC One. These media contents are consistent with the company's image, bolster Red Bull as being a lifestyle brand in these categories, and build awareness of Red Bull without necessarily promoting the product itself. An executive for Red Bull Media House North America remarked that the growth of digital media platforms had made it easier for brands to produce and distribute their own content and stressed that branded content was most effective when it is "authentic" and of high quality (Higgins, 2007). Third, marketing touch-points are becoming more personally relevant. The evolution of new kinds of media means that consumers are engaging more often in real-time conversations, particularly on social networks and other digital platforms. It is increasingly important for marketers to monitor conversations in social media and take immediate action, such as thanking a user for a positive comment or responding to a negative review with detailed explanation of how to address an issue or follow-up questions to learn more about what the consumer did not like. Some hotel chains, recognizing the importance of travel review sites (such as TripAdvisor), encourage satisfied guests to post comments online, while employing staff to follow and answer negative comments (Edelman and Salsberg, 2010).

6.4 Establish Online Presence

According to the Marketing Budgets Report by Oracle Marketing Cloud (2015), 77% of companies plan to increase their digital marketing budgets and 79% indicated that they are planning to increase their spending on digital marketing technologies. Some of the hottest digital channels for increased investment include content marketing (73%), search engine optimization (SEO) (60%), mobile marketing (57%), social media (56%), and paid search (54%), many

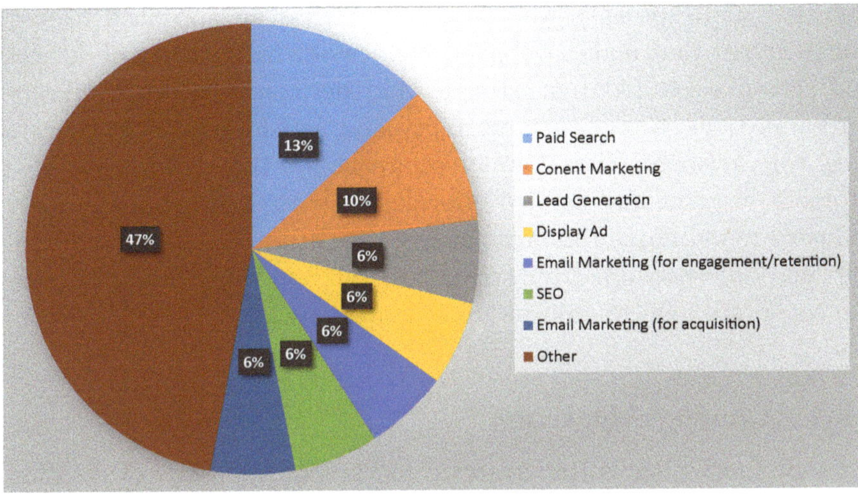

Figure 6.3: Digital Marketing Budget Split by Channel

Note: Other: Data management (5%), Display advertising (for engagement/retention) (5%), Social media (for acquisition) (4%), Affiliate marketing (4%), Marketing automation (4%), Marketing analytics (4%), Social media (for engagement/retention) (4%), Sales management (3%), Video advertising (3%), Webinars/Virtual events (3%), Mobile marketing (for acquisition) (3%), Lead management (3%), Mobile marketing (for engagement/retention) (3%).

Source: Oracle Marketing Cloud (2015)

of which will be discussed in this chapter. Figure 6.3 shows how digital marketing budget is split between these channels.

6.4.1 *Company Website and Corporate Blogs*

An organization's website is the core and foundation of its online presence. Two key attributes of an effective website are Integrated and Optimized. Integrated refers to the need to sync all elements of the digital presence, i.e., the official website and social channels. Optimized refers to the optimization of content and website for mobile devices. In digital marketing, content can be virtually in any form of written, auditory, or visual media. Marketers should focus on producing useful content that solves the problems of and adds value to the target audience. Another aspect of optimization is the web design. Responsive web design (RWD) is an approach to web design layout that makes webpages render well on a variety of devices and screen sizes. Content, design, and performance are necessary across all devices to ensure usability and satisfaction. A corporate blog can leverage on a human approach in the digital space to promote product and service information with relevant and timely industry news, human interest pieces, case studies, and other useful information. Switching from "corporate-talk" to "human-talk" allows the corporate blog to gain its distinct personality.

6.4.2 *Content Marketing*

Content marketing is the strategic marketing approach of creating and distributing valuable, relevant, and consistent content to attract and acquire a clearly defined audience with the objective of driving profitable customer action. Content marketing is critical for digital marketing because of two key reasons. First, content dominates consumers' online experiences. Interesting content is one of the main reasons people follow brands on social media. Overall, 50% of consumer time online is spent engaging with custom content (Johnson, 2013). Second, brands can build a long-term relationship with customers through consistent, valuable storytelling. 82% of consumers feel more positive about a company after reading custom content (Deflorian, 2018). By 2020, customer experience will overtake price and product as the key brand differentiator (Pemberton, 2018). Content marketing helps create the engaging experiences that customers seek and helps develop deeper connections between brands and consumers. Authenticity, relevance, and value are important aspects in digital content creation.

Content marketing comes in various forms, such as blogs, premium content, multimedia content, branded content, "smart" content, newsjacking, etc. The fear of missing out and the need for instant gratification dictate the dominance of short-form video content, especially on social media platforms. Live-streaming videos are also creating a niche for themselves, allowing brands to engage with viewers in real time by answering questions or asking for feedbacks.

Marketers can leverage on user-generated contents such as reviews and recommendations as they are genuine and appealing. Sometimes, brands can crowdsource contents as part of their campaign. For example, in 2014, Adidas tapped local artistic talent from Russia (St. Petersburg and Moscow) to promote the launch of the Adidas Originals AZ Flux — in an "Ode to the City". About 45 famous street artists from Moscow and St. Petersburg were commissioned to paint huge art objects (sneakers that were 3.6 meters long and weighed nearly 300 kilograms) in public areas like parks and main streets, and online voters were to select a winner from each city. Adidas hoped to harness the power of hometown pride through the work of local artists. The online campaign took off quickly. Social networks published over 1,500 photos of the work. The participating artists also actively encouraged their followers to vote for their sneaker design. During the voting period, the official website received over 1,150,000 visits. In all, 73,166 people participated in the vote. The ad campaign eventually reached over 143 million contacts (R3, 2014).

6.5 Social Channels

Beyond the organizations' core website, social channels form part of the other foundational element of the online presence. Social media are digital platforms that facilitate information sharing, user-created content, and collaboration across people. There is a significant increase in content consumption on social media platforms. Some of the more well-known social channels, i.e., Facebook, LinkedIn, Google Plus, Twitter, Instagram, Pinterest, and YouTube, are fundamentally changing the way we communicate, collaborate, consume, and create. Social media has revolutionized the ways organizations relate to the marketplace and society, creating a new world of possibilities and challenges for all aspects of the enterprise. Kaplan and Haenlein (2010) assign social media via social presence and self-presentation into six different groups: (1) collective projects (e.g., Wikipedia), (2) blogs and microblogs (e.g., Twitter), (3) content communities (e.g., YouTube or brand sites), (4) social networks (e.g., Facebook, MySpace, LinkedIn), (5) massively multiplayer online role-playing games (MMORPGs) (e.g., World of Warcraft), and (6) social virtual worlds (e.g., Second Life).

Peters and co-authors (2013) have proposed a framework of guidelines for social media metrics. Some of these guidelines can be applicable to both the measurement and the design of social media marketing efforts. For instance, the first guideline, *From Control to Influence*, emphasizes that organizations need to extract the essence of social media conversations and sentiments among the target audience and create relevant content. These contents should generate interests of not only the directly linked "fans" or "followers" but also the combined networks of organic reach. Because of the potential second- or third-degree reach through organic sharing, the marketing goal and measurement of influence should go beyond the initial direct network of followers. The second guideline, *From States to Processes & Means to Distributions*, indicates that the dynamics of the media network is more important than its performance in states. For instance, if a brand page is very popular in terms of total likes (a state), but the growth is slowing down, it may indicate a declining relevance. This process also depends on the heterogeneity of one's user base. For example, using data from Amazon, Sun (2012) shows empirically that a higher standard deviation of ratings improves a product's relative sales rank only when the average rating is below 4.1 (upon 5). The third guideline, *From Convergence to Divergence*, stresses the importance of considering the heterogeneity in user motives and consumption contexts. Certain brands may thrive on adversity in social media as differentiation increases and reinforces the identification of its core users. One example is Abercrombie & Fitch who seemed to face a social media crisis because of a quote from its Chief Executive Officer (CEO) stating that the brand is "exclusionary". The resulting adverse reaction toward the brand on social media by users may nevertheless improve the identification of its core brand users. Guideline four, *From Quantity to Quality*, highlights the intensity of nodes and links in a social network and different levels of engagement of various activities. With the rise of social ranking sites like Klout and PeerIndex, today's marketing communications are more and more steering toward a relatively small number of influential networkers. Traffic generated by prominent network users may carry more weight than others. In addition, the engagement levels of actors are associated with their motives, content, and social roles. Besides capturing frequency of keyword mentions of products/brands, the different types of interactions on social media reflect a corresponding level of engagement, e.g., a "like" has less value than a "comment" or compared to "sharing" content. These engagements cannot be "bought" but need consistent nurturing over time. Highly engaged fans (not necessarily in high quantity) are crucial in building sustained and authentic reach in social media. The fifth guideline, *Heisenberg Rule*, suggests that once marketers try to measure social media, its state and/or dynamics may be altered. For example, KloutScore influence score is measured by the topic, the format and content, and how audience react to the content, while EdgeRank is a composite of affinity, weight, and time decay. As these underlying rules are transparent, marketers need to be aware that individual users may manipulate the social interactions to generate the desired outcome. The sixth guideline, the "Ying & Yang" rule posits that sustainable growth of a fan base can only be assessed by qualified growth combining metrics on growth in size and the quality of engagement levels. For instance, a KloutScore can only work over time when balanced by a metric that punishes "fake" engagement and impact. Social networks are always changing in size, composition, usage levels, and structure. For example, heavy users of social media tend to adopt earlier than people with lower usage. Accordingly, average usage time may eventually go down over time as more people join these media. Lastly, guideline seven, *From General to Specific*. Consumers may use specific platforms for specific actions, e.g., Twitter to complain (as they desire a fast company reaction), Facebook to boast about successful purchases (as it only goes out to close friends), and Instagram to combine brand visuals. In the meantime, users are active across social media and subsequently one can regularly observe spillovers, e.g., from Twitter to Facebook activity and vice versa (see Table 6.4).

Table 6.4: Selected Guidelines for Designing Social Media Metrics

Guidelines	Recommendations
Guideline #1: From Control to Influence – Brand communication shifts from a patriarchal to a participatory paradigm	• Sustained reach cannot be bought like in traditional media. • To generate relevant content that fits the motivational structure of the brand's audience, managers need to build up the capabilities to listen and respond to not only the directly linked consumers but also the combined networks of organic reach of target audience.
Guideline #2: From States to Processes and Means to Distributions – The importance of dynamics	• Growth or decline in numbers, a relevant signal. • Traffic generated by prominent posts.
Guideline #3: From Convergence to Divergence – Account for divergence and heterogeneity	• Adversity in social media as differentiation increases and reinforces the identification of its core users. • Converged assessment will be senseless without accounting for the motives of the targeted network population and the context of evaluation.
Guideline #4: From Quantity to Quality – Intensity of nodes and links and different levels of engagement	• The engagement levels of actors that are tied to motives, content, and social roles. • Different types of interactions (e.g., likes, comments, or shares) reflect different levels of engagement.
Guideline #5: Heisenberg Rule – The measurement of social scores may alter the state and/or dynamics	• The underlying rules of social scores like KloutScore, EdgeRank are transparent. • Social scores may be games and the dynamics may be distorted.

Table 6.4: (*Continued*)

Guidelines	Recommendations
Guideline #6: Ying & Yang Rule – Growth in numbers and the quality of engagement levels	• Growth in size and the quality of engagement levels. • Social networks are always changing in size, composition, usage levels, and structure.
Guideline #7: From General to Specific – Specific platforms for certain actions and spillovers	• Consumers may use specific platforms for specific actions. • Active users may observe spillovers from one platform to another.

Source: Adapted from Peters *et al.* (2013).

6.5.1 *Paid Social Promotion*

Apart from the typical organic social marketing efforts, many of the social media platforms/channels have paid advertising tools to assist organizations to promote their branded social pages to other users on social media platforms. Facebook has the most established advertising system among the social media platforms discussed. Facebook advertising adopts the pay for performance or pay-per-click business model, as well as cost per thousand (CPM) impression model. The Facebook Ads are presented on a user's Facebook interface (typically located on the top right-hand corner). Facebook sells its ads to its advertisers via targeted user segments by setting parameters that include user profile demographics, location, relationship status, interests, comments, likes, and other user activities captured by its algorithm (EdgeRank). Once a target user logs into his/her Facebook page, Facebook will present the advertisers' ads that match the users' profile and/or other parameters previously entered by the advertiser.

Another key advantage of paid social promotion is that many of their ad formats allow them to embed the users' social contacts to leverage on word-of-mouth and social proof, e.g., their sponsored stories ads. Users would be able to see which contacts have liked, shared, or commented on a specific sponsored post or event.

6.5.2 *Search Engine Optimization*

Search engines (SEs) send web-scraping spider — an internet-based tool to "crawl" the web to find new documents, such as webpages, typically by following hyperlinks from websites already in their database. Then SEs index the content (text, code) in these documents by adding it to their databases and then periodically updating this content. Search engines search their own databases when a user enters a keyword (a particular term, phrase, or text) in a search to find related documents instead of searching webpages in real time. Search engines rank the resulting documents using an algorithm (mathematical formula) by assigning various weights and ranking factors.

Google dominates the SE market with close to 80% share on desktop and 96% share on mobile (Desjardins & Visual Capitalist, 2018). Figures 6.4(a) and 6.4(b) show the SE market share on desktop and mobile, respectively.

Search Engine Optimization refers to the process of "optimizing" both the on-page and off-page ranking factors in order to achieve high search engine rankings for targeted search terms (keywords). It refers to the "industry" that has been created regarding using keyword searching as a means of increasing relevant traffic to a website. Providing the conversion rate is good, more relevant web traffic can generate more leads and sales. Typically, search engines will not openly share about their SE algorithm and most SE algorithms are constantly changed, tweaked, and updated. Such algorithms also vary by SE — some give more weight to on-page factors and some to link popularity. In the meantime, websites and documents being searched, crawled, and indexed are also constantly changing, which poses lots of challenges to SEO professionals. In practice, reverse engineering, research, and experiments give SEO agencies and professionals a

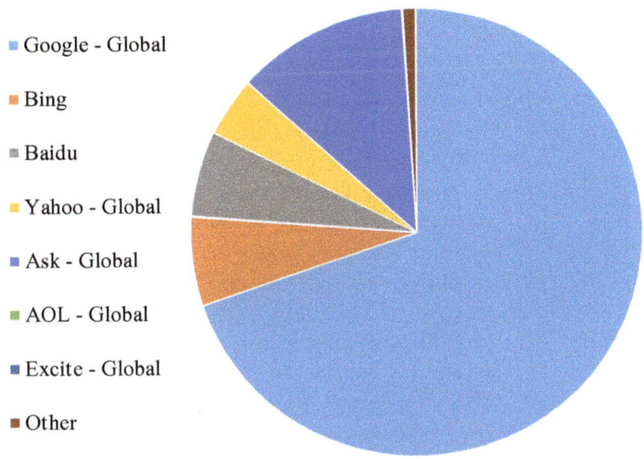

(a)

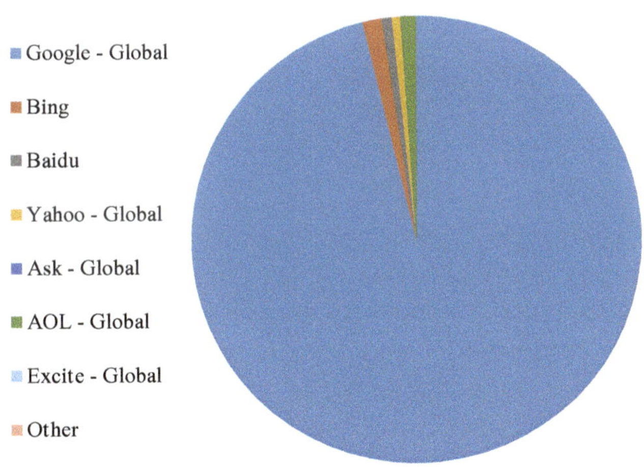

(b)

Figure 6.4: (a) SE Market Share on Desktop, (b) SE Market Share on Mobile

"good enough" idea of the major factors and approximate weight assignments. SEO is not about gaming SE's search result ranking algorithm — it is about optimizing content that is fresh, original, and targeted for search engines. Two key optimizing tools are keywords and link building (see Table 6.5).

Table 6.5: Key Factors That Affect the Visibility of a Website

Key Factors	How It Affects the Visibility of a Website
Keyword optimization	The key principle is that keywords need to resonate with your target audience. Thus, you need to research into who your target audience is/are, where they are actively interacting in the digital space, what their current wants or needs are, and how they are "talking" about these issues. Your content should be developed for quality, freshness, and relevance and it should have these keywords embedded within that content. You should not write up the keywords and then build content around them. Content first, keyword second. Keywords need to permeate through your website design even in the URLs and Title/Header Tags.
Link building	Content marketing via blogs and engaging in content sharing on your social media platforms and channels are important link-building efforts.

6.6 Case Study — Digitalization in India

India is blessed with a few advantages: a youth population (aged 10–25) with over 150 million people entering the workforce by 2035, fastest growing trillion-dollar economy in the world, and strength in technology revolution. Undoubtedly, India is leapfrogging a few stages to embrace the power of digitization. Demonetization, a bold move by the government of India, has been slowly changing the transaction patterns of the nation.

Devices are becoming cheaper and mobile phones are becoming pivotal in people's daily lives. The widespread roll-out and adoption of 4G technology and the push behind digital literacy beyond major cities has opened up new opportunities. Be it ordering a cab, booking a dinner table at a restaurant, or shopping, consumers are interacting with various brands across industries via digital media. Therefore, marketers now have to redirect their marketing efforts in creating a micropersonalized, contextual, and enriching experience across various touch-points. According to the Confederation of Indian

Industry and KPMG (2017), marketers need to better connect with the customers in India via the following strategies:

- **Mobile first:** Mobile strategy is no longer a choice but a necessity in today's marketing environment. From tablets to smartphones, from smart watches to virtual reality (VR) headgears, consumers strive for more immersive mobile solutions. For instance, agri-focused organizations in India are empowering millions of farmers with mobility-led solutions, which not only cater to the farming needs but also go beyond that to connect with the farmers as an individual.
- **Omni-channel:** IoT is making the omni-channel experience more immersive and unified. Each connected device signifies a new channel which helps marketers generate customer data, such as location and consumption behaviors. Marketing organizations can use it to draw insights on their entire buying journey.
- **Content is king:** Businesses should invest in content creation and leverage on content optimization tools and data analytics to increase the efficiencies and reduce the cost of customer acquisition. Video is gaining popularity over static visuals and texts. Video analytics can help to gain insights on what type of content sells and how to increase the reach at each stage of the marketing funnel to constantly revise and optimize their efforts.
- **Mastering the art and science of analytics:** A seamless flow of data between sales, marketing, and service functions is key for unlocking the potential of digitization. Taking data insights into consideration while planning for campaigns can build a solid foundation for a successful digital strategy.

Source: Confederation of Indian Industry and KPMG (2017).

6.7 Conclusion

The pace of adoption of digital channels is pushing marketers to deliver exceptional seamless experience. Instead of providing a promotional message to customers, marketers are expected to deliver consistent experience throughout the customer journey across all channels, be it digital or physical. As consumers become less loyal to brands and more inclined to research using search engines and social channels, companies have to align themselves with this new generation of customers' needs.

Application of big data and advanced analytics can play a critical role in precision marketing, when marketers integrate and exploit multiple data sources to understand the specific wants, needs, interests, and behavioral patterns of particular user segments. With the help of customer insights, marketers can identify, target, retain, cross-sell, and upsell to potential leads and existing customers by providing relevant and unparalleled customer experience, instead of flooding communication channels with mass-reach messages.

References

Coleman, L. D. (2017). Inside the Wild Tech Plans of Alibaba to Dominate 11.11, *Forbes*. Retrieved from https://www.forbes.com/sites/laurencoleman/2017/10/31/inside-the-wild-tech-plans-of-alibaba-to-dominate-11-11/#2e6fab22dc95.

Confederation of Indian Industry and KPMG (2017). Digital — The New Normal of Marketing. Retrieved from https://assets.kpmg.com/content/dam/kpmg/be/pdf/Markets/digital-the-new-normal-marketing.pdf.

Corcoran, S. (2009). Defining Earned, Owned, and Paid Media (accessed June 12, 2012). Retrieved from http://blogs.forrester.com/interactive_marketing/2009/12/defining-earned-ownedand-paid-media.html.

Deflorian, A. (2018). Loyal to Content: Thinking Beyond Punch Cards for Loyalty Programs, Forbes Agency Council. Retrieved from https://www.forbes.com/sites/forbesagencycouncil/2018/09/27/loyal-to-content-thinking-beyond-punch-cards-for-loyalty-programs/#5178f7644666.

Desjardins J. & Visual Capitalist (2018). How Google Retains More Than 90% of Market Share, *Business Insider*. Retrieved from https://www.businessinsider.com/how-google-retains-more-than-90-of-market-share-2018-4?IR=T.

Edelman, D. & Salsberg, B. (2010). Beyond Paid Media: Marketing's New Vocabulary, McKinsey Quarterly. Retrieved from https://www.mckinsey.com/business-functions/marketing-and-sales/our-insights/beyond-paid-media-marketings-new-vocabulary.

Goodall, D. (2009). Owned, Bought, and Earned Media. *All That Is Good*, (March 2), (accessed June 12, 2012). Retrieved from http://danielgoodall.com/2009/03/02/owned-bought-andearned-media/.

Higgins, M. (2007). Red Bull's Headlong Frozen Dash Is a Crash Course in Marketing. *The New York Times*. Retrieved 2018-01-28.

Johnson, S. (2013).*13 Spooky Stats to Scare Your Boss Into Better Marketing.* HubSpot, Inc.

Kaplan, A. & Haenlein, M. (2010). Users of the World, Unite! The Challenges and Opportunities of Social Media, *Business Horizons*, 53(1), 59–68.

Kaufman, I. & Horton, C. (2015). *Digital Marketing: Integrating Strategy and Tactics with Values. A Guide Book for Executives, Managers & Students.* Routledge, Taylor & Francis Group, New York.

Oracle Marketing Cloud (2015). Oracle Marketing Cloud and Econsultancy Launch Marketing Budgets Report 2015. Retrieved from https://blogs.oracle.com/marketingcloud/oracle-marketing-cloud-and-econsultancy-launch-marketing-budgets-report-2015.

Pemberton, C. (2018). Key Findings from the Gartner Customer Experience Survey, Gartner. Retrieved from https://www.gartner.com/smarterwithgartner/key-findings-from-the-gartner-customer-experience-survey/.

Peters, K., Chen, Y., Kaplan, M. A, Ognibeni, B., & Pauwels, K. (2013). Social Media Metrics — A Framework and Guidelines for Managing Social Media, *Journal of Interactive Marketing*, 27(4), 281–298.

R3 (2014). *Integration 40.* Retrieved from http://r3integration40.com/.

Soat, M. (2016). Google VP on the Marketer's New Normal, AMA Annual. Retrieved from https://www.ama.org/publications/eNewsletters/Marketing-News-Weekly/Pages/google-marketers-new-normal.aspx.

SparkToro (2018). New Data: How Google's Organic & Paid CTRs Have Changed 2015–2018. Retrieved from https://sparktoro.com/blog/new-data-how-googles-organic-paid-ctrs-have-changed-2015-2018/.

Stephen, A.T. & Galak, J. (2012). The Effects of Traditional and Social Earned Media on Sales: A Study of a Microlending Marketplace, *Journal of Marketing Research*, 49(5), 624–639.

Sun, M. (2012). How Does the Variance of Product Ratings Matter? *Management Science*, 58(4), 696–707.

Chapter 7

From E-commerce to New Retail

7.1 Introduction

With the birth of the computer network, one of the industries most affected and undergoing tremendous changes is the retail industry. The discussion that "Internet kills retailing" or "retailing is dead" has been making headlines on various business magazines. Then surprisingly, as time goes by, we see somehow the two move into an era when the two former opponents co-exist in harmony. Those conventional brands were busy moving online almost in a panic, while those online-born brands, amazingly, went brick-and-mortar. Online to offline, offline to online, online and offline, how did it all begin...?

7.2 A History of Modern Retail

From trading camels (still ongoing) to sending a pack of rice ordered by a rice dispenser, the emergence of organized retail started initially through posting catalogs at the end of the 1800s in the US. The population was scattered and access to stores was very limited. Catalogs were sent to the consumer's house and it demonstrated a brilliant distribution strategy: placing the store in front of your customer's front door, a fantastic manifestation of the true adage in retail industry, "Location, location and location". This business model was initiated by Richard Sears who partnered with Alvah Roebuck in 1886 (Sears, 2003) to form the basic system of classic American retailer Sears. The Sears catalog eventually increased to more than 500 pages from the cradle one rocks their baby to the coffin they would be buried in. In the 1920s, when the population migrated from farms to small towns, Sears followed the consumers by opening shops in those small towns. There they had a multichannel distribution strategy, both the store and catalog. Then they were able to sell fair quality products with a credible price and started to position themselves as the go-to stores for the growing middle class who were not served by the department stores in the city center.

The department stores, at that time, were called "cathedrals of shopping". They were the daylong outing places for families. Initially, families were there because of the breadth of offerings carried by these stores. Later, they were there because of the sponsored entertainment, children's event, fashion show, restaurants, and more. In the late 1980s, Kendals (now House of Fraser) and Harrods opened their first stores in Manchester and London (House of Fraser, 2005). In Paris, the cousins, Théophile Bader and Alphonse Kahn, opened their shop at the corner of rue La Fayette. Over the next decade, they bought up surrounding properties and expanded Lafayette's products to various categories, including fashion, home goods, gifts, toys, etc. (Galeries Lafayette, 2004). Burgeoning retail industries began to emerge in Asia, the Middle East, and remaining parts of the world. In this period, a common view in the retail industry goes as the line from the movie *Field of Dreams,* "if you build it, they will come".

In the 1970s, post-World War II, explosive growth was across almost all industries in the US, including retailing and other consumer-related industries. Such growth was encouraged by the massive development of a nation's communication, transportation, distribution, and marketing infrastructure (Lewis and Dart, 2014). This unprecedented economic prosperity brought a huge jump to the American standard of living. It is a cycle of affluent job opportunities providing discretionary income, thus increased demand which requires more production and more jobs to offer. Two phrases were coined to describe the market dynamics in this booming economy "mass market" and "mass marketing". This was a golden age for innovation, advertising, and branding. Big brands were expanding and new brands were sprouting daily, such as Procter & Gamble, Coca-Cola, Levi Strauss & Co., General Motors, Ford, Timex, Playtex, Anheuser-Busch, Gap, General Mills, R. J. Reynolds, and more. New formats of the retail model such as category killers, specialty chains, and discounters started to emerge. Gap began its meteoric rise from the little shops in San Francisco selling Levi's jeans to the largest private-label-branded apparel retailer in the world. Esprit has closed its operations in the US, but remains a strong international brand.

With more competitors entering the consumer-related industries, markets became saturated in the 1980s. This saturation basically exists in all industries, such as retail, automobiles, airlines, consumer packaged goods, services, travel, and entertainment. In an era of multibrands in a single category, consumers asked for a compelling reason to buy from one rather than another. Competitive congestion forced companies to innovate to differentiate from their competitors to avoid blooded competition over price war. It also shifts the market power from the producers to the consumers.

Between 1980 and 2010, this penultimate phase of one of the economic shifts in history is called "an enormous global power shift from producers to consumers" by Professor Rosabeth Moss Kanter of the Harvard Business School. In a research by the International Council of Shopping Centers, the massively growing shopping space reached 7.5 billion square feet in 2013 in the US, which was more than twice as fast as the rate of population growth (Kercheval, 2013). Under a global context, United Kingdom is a runner-up with three square feet per capita. Outlets or discounters, such as Walmart, Costco, Home Depot, Lowe's Barnes&Noble, Best Buy, and Toys "R" Us, are one of the main factors driving the explosive growth of the shopping space. They also provide more and cheaper access of products to consumers.

Then finally, the Internet and all the new technological innovation seemed to be evolving in an orderly pace, and then such peaceful evolution turned out to be a full-blown revolution. Amazon accelerated its expansion into multiple retail and distribution platforms, including a huge investment of an early bet on cloud technology. eBay woke up and changed their auction model after the Amazon testimony. PayPal and other businesses rapidly expanded to strengthen their ties with consumers. Google began to transform the way retailers used to reach their consumers. Social networking sites such as Facebook have various communities, powerful enough to influence consumer perception toward brands and social and economic issues. We are in an era where more power is accruing to the consumer.

7.3 Software Eats Retail

Marc Andreessen, one of Silicon Valley's most influential investment gurus and one-half partner of the venture capital powerhouse Andreessen Horowitz, which has interests in an array of Internet technology firms, including Facebook, Airbnb, Twitter, and Pinterest, stated in an interview with Sarah Lacy (Lacy, 2013):

> "Retail guys are going to go out of business and ecommerce will become the place everyone buys.... Retail chains are a fundamentally implausible economic structure if there's a viable alternative. You combine the fixed cost of real estate with inventory, and it puts every retailer in a highly leveraged position. Few can survive a decline of 20 to 30% in revenues. It just doesn't make any sense for all this stuff to sit on shelves. There is fundamentally a better model."

Andreessen's comment sparked a firestorm of debate. Some felt his argument to be a bluster with obvious investment purpose on the technology firms he has interest in. They also decried pure online retail as a loser in the real world being propped up by venture capital. Others took a different view and pointed to the disastrous impact of technology toward brick-and-mortar shops, such as Netflix replacing Blockbuster in the video rental market or the online travel options to the travel industry.

In 2012, Walmart's revenues were 16 times those of Amazon, making it easy to dismiss the e-retailer as a profitless nuisance. Only six years later, Amazon's revenue shrunk to less than three times. Amazon's EBITDA was US$15.6 billion compared to US$28 billion of Walmart. What probably was more worrying in that year was the unconceivable fact that Walmart suffered its first sales decline since it went public 45 years ago. For Walmart, the bitterest pill to swallow also came in the same year when Amazon unseated Walmart as the world's most valuable retailer. Amazon's market capitalization had exceeded that of Walmart by US$20 billion.

In June 2016, Walmart formed an alliance with China's second largest business-to-consumer e-commerce company, JD.com. It demonstrated its efforts to take a share of China's burgeoning e-market. That same year in August, Walmart acquired a US-based e-commerce one-year-old start up Jet.com at US$3 billion. Marc Lore, the founder of Jet.com, sold his first start-up company to Amazon in 2010 and believed that with advanced analytic technology Jet would outprice Amazon. The purchase not only brought Walmart a pricing model but also Marc, who would be at the helm of Jet. Despite Walmart's efforts to enhance its online capability, its customer base might not line up well with the profile of the online shopper. Its core customer is not very tech-savvy. The customer who used to pave Walmart's unprecedented dominance in retail had become one of its greatest barriers. Is it time to bid farewell to Bentonville, the town that would not be known if it were not for Walmart?

7.4 The Online Behemoth

When Macy's CEO Terry Lundgren commented regarding the threats posted by Amazon in the apparel industry, he believed Amazon is not well prepared for the harsh reality of online apparel orders. The challenge according to Lundgren was to deal with the returned orders. While Lundgren was making such comments, Amazon had a very different focus. Amazon's CEO Jeff Bezos was holding a by-invitation-only conference on robotics, artificial intelligence, and space exploration with the world's top experts in the field. So while Lundgren was pondering the complexity of inventory and store operation in the apparel industry, Bezos was probably discussing commercialized space trip, robotic advisory, and more. As agreed by its industrial rivals, using conventional formula to strategize Amazon had hardly worked out.

Prime is at the center to drive Amazon's profit and sustain growth. Prime is a subscription program. With an annual subscription fee at US$99, the members enjoy a free two-day shipping and access to a range of other services and programs offered by Amazon. It turns out that Prime members are not only more tech savvy and engaged customers but also more profitable customers. While Amazon's non-Prime members spent on average US$625 on Amazon in 2015, it was US$1,500 for its Prime members! In addition, Prime members had an amazingly low churn rate. Overall, 75% of 30-day trial memberships ended up with a full-year membership renewal and 96% membership of the second-year paid members was renewed for a third year.

Amazon hardly thinks of short-term products and services like most of the retailers. For example, when building its Echo device, it released the Application Programming Interface to allow other developers to build this feature into their own products. When innovating, Amazon brings values to not only its consumers but also other business partners. The fact that Amazon often innovates on a different strategic domain makes it almost impossible for its competitors to anticipate its next move or follow its step. Its capacity to run so many parallel innovation projects at a time across different fields is unprecedented. In addition, its high threshold for failure within innovation is usual, something Amazon wears like a badge of honor.

7.5 The Asian Beast

Every November 11, the Chinese consumers celebrate the so-called Singles' Day as the date 11/11 or four singles. This is a day for young and attached people to get together for parties or other social events. As the celebration grew popular, the retailers used it to target consumers with promotions. The e-commerce giant Alibaba is leading the change. On November 11, 2016, Alibaba sold a startling US$5 billion merchandise in the first hour of the Singles' Day. Yes, the sales in that hour almost doubled the sales of all retailers in America on Black Friday! Alibaba finished the Singles' Day by selling more than US$20 billion, two times that of eBay's annual revenue in 2015. It was also a 40% increase on a year-over-year comparison.

In 2017, Alibaba sold an astounding US$541 billion worth of commodities across its platform (Alibaba's Revenue 2017, 2017). More noteworthy is the fact that while most of the technology firms are criticized for their heavy top line and thick bottom line, Alibaba's gross profit margin was more than 36%, larger than the profit of both Amazon and eBay combined.

In 2015, China surpassed the US to become the world's largest e-commerce market. By 2015, China had twice as many Internet users as the US and 89% of those users can use mobile phones to access the e-commerce platforms. Considering the fact that only half of the population in China is digitally connected, one can imagine the scale of the future e-commerce economy in China when it reaches near-full connectivity.

7.6 The New Retailer

In a letter to his shareholders in October 2016, Jack Ma coined the term "New Retailer", his vision of the future for retailers. He posited "Pure e-commerce will be reduced to a traditional business and replaced by the concept of New Retail — the integration of online, offline, logistics and data across a single value chain." In a video (Alibaba's 'New Retail' Explained, 2018) by Alibaba to demonstrate Ma's vision of the new retailer, it is explained that the future is not about total dominance of e-commerce, rather it is the digitization of any commerce. There is no such "either-or" retailer but blending the best of both in-shop and online experiences.

To move this "New Retail" concept to the forefront, 64 Hema stores in 14 cities (Alibaba Says New Retail Strategy Is Paying Off as Hema Shopper Data Shows Bigger Average Spending, 2018) were set up as a test ground in 2016. Hema supermarket allows shoppers to order their groceries online and promises to deliver them within 30 minutes for those who live within three kilometers. Alternatively, the consumers can also shop in the store and have their goods delivered at their doorstep. Starving diners can also pick fresh seafood and let the chefs present the cate. A Hema store, which has been open for more than one and a half years, on average generates a daily sales of US$116.50. A Hema store can double the profit margin of traditional grocery stores. The integrated online and offline shopping business models increase consumers' shopping opportunities. A consumer who purchases both online and offline on average spends US$82 monthly compared to $43 by a consumer who only purchases from a single channel.

In an effort to transform a traditional neighborhood mom-and-pop store, Alibaba built an app that provides a list of products most likely to be ordered in a particular neighborhood based on their past customer analytics in that neighborhood. In addition, they also tie these businesses with a central warehouse and logistic system. In this way, the mom-and-pop stores benefit from less lost sales due to stock-out, use their limited space in the most profitable way, and save the time and effort to deal with multiple suppliers as they used to do.

Then, how about the malls? One reality of the mall-shopping experience is that you often find the right pants, but not the right size or color in stock. Who wants to leave the shop empty handed? The New Retail malls will be equipped with "virtual shelves". In case your preferred color or size is out of stock, you can simply secure a stock by keying in the specs and have it delivered. Then comes the lady's powder room. Ladies will appreciate it when they do not have to rubber their lips to try different colors. With the "magic mirror" on the wall, one can experiment virtually the most daring color one would never pick up. As commented by Alibaba's CEO Daniel Zhang, the firm believes in the value created by the brick-and-mortar stores. However, these offline physical stores need to be upgraded and digitized for their operation to deliver a perfect customer experience.

The last manifestation of the New Retail concept we cannot afford to skip is the new car selling model. Anyone who has the experience of buying a car knows how exhausting it is to visit different car dealers to test-drive different car models. With Alibaba's unmanned auto vending machine, you can simply book the car model and drive it away up to three days. Be it sending the kids to school or a family picnic in the weekend, gone are those tremendous pressures from the salesman. Once after trying under a pressure-free environment, you can decide a dealer to buy the car or buy it through Alibaba.

In a global summit of consumer goods forums, Zhang mentioned that the core value that allows Alibaba to grow to today's scale is its culture to embrace changes. "We evolved our business first based on the needs of customers. Secondly, from a business perspective, to enable our business partners to do business successfully on our platform" said Zhang (Wan, 2018). No wonder changes happen faster in China's retail sector than elsewhere, as asserted by a quote from Jack Ma "The key to save traditional retail is to forget about tradition all together".

The New Retail model versioned and currently rolled out in China essentially means creating a holistic consumer experience through integrated channels with seamless operation. With this achieved, the customer will establish certain emotional affiliation with the brand. Such emotional affiliation can be referred to as neurological connectivity. The seamless melding of the online and offline channel can be referred to as omni-channel. These two concepts will be discussed in the following sections.

7.7 Neurological Connectivity

Neurological connectivity is achieved when a retailer, brand, or service creates a strong psychological and emotional response that operates on a subconscious level for the consumer in a way that is typically neither readily understood nor necessarily recognized by the consumer (Stephens, 2017). Studies in neuroscience show that when people have an elevated experience, their brain releases different chemicals. One in particular is dopamine, which leads to feelings of euphoria, self-satisfaction, and well-being. Consumer connectivity is thus not about physical connection through products or services alone. In other words, a brand or store has a neurological connection with its customer; so when the customer pays a visit to the store, it is as if he or she is dropping in to visit a good friend.

Successful models of neuroconnectivity target to drive the entire experience of four things: the dopamine rush in anticipation of shopping, compelling the consumer to visit the store, the joy of the actual shopping experience itself, and the final satisfaction of consuming or using the product. An *MIT Sloan Management Review* by Coimbatore Krishnarao Prahalad and Venkat Ramaswamy stated that "Value creation is defined by the experience of a specific consumer, at a specific point in time and location, in the context of a specific event. The experience space is conceptually distinct from that of the product space, which is the conventional focus of innovation. In the experience space, the individual consumer is central, and an event triggers a co-creation experience. The events have a context in space and time, and the involvement of the individual influences that experience. The personal meaning derived from the co-creation experience is what determines the value to the individual." Simply put, when a consumer visits a retailer, he or she has a unique enjoyable experience that cannot be repeated outside of this relationship.

7.8 Omni-channel

Omni-channel is a holistic approach to sales that seeks to provide the customer with a seamless shopping experience whether the customer is shopping online (from a desktop or mobile device, by telephone) or offline in a brick-and-mortar store. The difference between omni-channel and multichannel is that an omni-channel emphasizes the integration of all channels in creating customer experience. The entire focus of an omni-channel is to neurologically connect with their customers to create incredible experience through personalized products and superior services. The multichannel approach, on the other hand, focuses on using different channels to reach customers of different segments. The power of the model will compel more pure e-commerce sites to set up physical stores to expand their e-commerce and mobile capabilities. When visiting an enticing website, the consumer may be looking to feel the same engrossing experience in the physical store, and vice versa.

Most of the brands that apply the omni-channel concept occupy 3,500–5,000 square foot physical stores. Such a set up provides a casual, more intimate, and less frustrating shopping environment as compared to a big department store or big-box models. It caters to shoppers who desire a relaxed shopping experience and those who want a quick entry and exit. This also ensures a single dedicated space for all product assortments under the brand rather than having different models or versions of the product scattered over different department stores. Employees serving in these stores are usually employed from the customer base, people who love the brands and could share immediate feedback with the customers. With technological advancement, virtual reality (VR), augmented reality (AR) or artificial intelligence (AI) will be more widely exploited by such stores to provide holistic customer experience with the aim of achieving neurological connection with customers.

Omni-channel is defined as an integration of all channels of a brand that provides the customer with a holistic and seamless customer experience. Unified and cohesive product proposition and assortments should be available in all channels. Whether the customer is shopping online or offline, the target is to create a neurological connection with the customer.

7.9 The End of the Beginning

E-commerce has totally re-programmed us. The online shopping experience rewrites our expectation toward offline shopping. How many times have you poked a screen that is not a touch screen? How many times in a store with huge assortments have you wondered where the filter function is? With the brick-and-mortar retail threatened by e-commerce, we are wondering when the expansion in online business will regress to the growth rate of retail market as a whole? When the astounding double-digit growth rate of online business will return to the 3–5% of growth that the conventional retailer is accustomed to seeing? Say it differently, what does the future hold for the conventional retailer? Is Alibaba's New Retail a cure? In other words, what will the future of retail look like?

No matter how disruptive the future technologies can strike or shake the business world, retail as a consumer-facing industry should recognize the power and significance of the convergence of art and science of retailing. This convergence will drive brands, retailers, or even the whole consumer-facing industries.

References

Alibaba Says New Retail Strategy Is Paying off as Hema Shopper Data Shows Bigger Average Spending. (2018). Retrieved from scmp.com:

https://www.scmp.com/tech/enterprises/article/2164651/alibaba-says-new-retail-strategy-paying-hema-shopper-data-shows.

Alibaba's 'New Retail' Explained. (2018). Retrieved from Alizila: https://www.alizila.com/video/alibabas-new-retail-explained/.

Alibaba's Revenue 2017. (2017). Retrieved from China.com: https://economy.china.com/news/11173316/20170519/30543228_all.html.

Galeries Lafayette. (2004). Retrieved from wikipedia: https://en.wikipedia.org/wiki/Galeries_Lafayette.

House of Fraser. (2005). Retrieved from Wikipedia: https://en.wikipedia.org/wiki/House_of_Fraser.

Kercheval, M. P. (2013). *Industry Updates.* Fairmont Dallas: ICSC.

Lacy, S. (2013, January 30). *Andreessen Predicts the Death of Traditional Retail. Yes: Absolute Death.* Retrieved from PanDo: https://pando.com/2013/01/30/andreessen-predicts-the-death-of-traditional-retail-yes-absolute-death/.

Lewis, R. & Dart, M. (2014). Demand Creation in a Marketing-driven Economy. In R. Lewis and M. Dart, *The New Rules of Retail* (p. 238). New York: St. Martin's Press.

Sears. (2003). Retrieved from Wikipedia: https://en.wikipedia.org/wiki/Sears

Stephens, D. (2017). *Reengineering Retail.* Group West.

Wan, L. (2018). *Alibaba CEO New Retail Means Providing Value Maximising Data and Embracing Change.* Retrieved from foodnavigator: https://www.foodnavigator-asia.com/Article/2018/07/17/Alibaba-CEO-New-retail-means-providing-value-maximising-data-and-embracing-change.

Chapter 8

Web Analytics

Written jointly with Wang Yue

8.1 What Is Web Analytics?

Web analytics is a way of learning how users interact with the website by recording the data and transforming the data into behaviorally interpretable insights. The most valuable information web analytics tools collect is how pages are viewed by a user, namely, what have been viewed in what order. In essence, web analytics is able to restore a visitor's journey through different webpages on a website.

In the past decade, websites have really grown to be an accountable channel for most of the companies as other channels, such as phone, retail, dealer, etc. Firms are in demand to justify the investment poured into web leadership. The web analytics tools have matured since the mid-2000s. Much of the use of web analytics has been in the area of online marketing, a field concerned with connecting the firm with its customers. Web analytics allows the markcters to measure the effectiveness of their marketing initiatives through various metrics, such as the number of people who reach the website and go on to buy things.

8.2 Log File Analysis and Page-tagging Approach

Some basic understanding of how web analytics tools work helps to comprehend the limitation of such tools as well as interpretation of the data and insights based on such data. There are two basic approaches: log file analysis and page tagging. Log files keep track of what pages are loaded upon the requests received by a server. It allows you to deep dive into the data, but is more challenging to use. On the other hand, page tagging is less accurate, but usually involves fewer obstacles to use as a web analytics tool. Within the category of page tagging, there are tools that enable you to study how users move from page to page and tools for analyzing what they are actually doing on each page.

While log file analysis is out of the scope of discussion of this book, the concept delivered here transfers well. Web servers keep visit records. When a request is sent to a server that hosts a website, whether such request is sent from a web browser or a search engine, it gets recorded in the server's log. Such recording does not require the user to enable JavaScript, which is a weakness of page tagging. Why is log file analysis not widely used? Partially, it is because page-tagging tools are relatively well marketed as web analytics tools. In addition, page-tagging tools are much easier to use and maintain. The log file analysis tool is expensive to acquire, also the size of the log file can be huge and challenging for an IT department to share.

The page-tagging method works by allowing a small piece of JavaScript code to all the webpages one wants to track activity of. When a user's web browser loads a webpage, it starts to open the Hypertext Markup Language (HTML) file and interpret the code on the webpage as shown in Figure 8.1. The JavaScript asks the user's browser to send the following information:

- What page has just been loaded
- When was it loaded
- Whether the user comes from a search engine or a link from the link of another website
- IP address
- User details such as the browser type, operating system, screen resolution, etc.

Often it also puts a cookie on the user's computer to let the analytics tool know if a particular user's data have been recorded before. Two types of page-tagging tools are available to track user

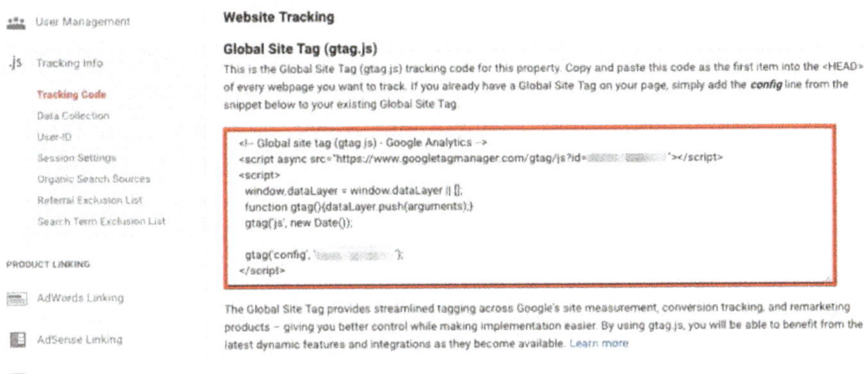

Figure 8.1: An Example of Google Analytics' Tracking Code

activity. A Google Analytics type of tool allows one to study the user's move from one page to another page. Another category of the page-tagging tool such as TeaLeaf, Crazy Egg, and ClickTale enables research of user's activity on a particular page. Such analytics is often referred to as click analytics.

8.3 Understand Both the *What* and the *Why*

Suppose while shopping in a supermarket, if one did not purchase anything, the supermarket's manager would probably not know that you were there. If you have purchased something, with a membership card, they may know who has purchased the product at what price and time. They may also infer the product inventory at home based on your purchase history.

Visiting a website is rather a different proposition from the lens of data collection. During one's journey on a website, one leaves behind significant amount of data, irrespective of whether purchase takes place. The website knows every 'aisle' one walked, the label one stayed reading, and everything one puts in a cart and discarded eventually. Hence, even if one visited and left the website, it still knows everything one did and the sequence in which it was done. We can then see how massively advantaged the web is in terms of its ability to collect data without violating the core of customer privacy.

However, no matter what analytical tool one is using, the collected data will only enable you to understand *what* happened but not *why* it happened. We have clicks, pages viewed, bounce rates, paths, abandonment rates, etc. We miss a critical facet: Why did the user click? Why did they abandon the shopping cart? If the bounce rate of a particular page is so high, is it because they are so confused about the content thus leaving the website or have they found the information they need? Apparently, bounce rate as a web analytics metrics only tells the percentage of visitors exiting the website after viewing the page. This is a *what* facet but not the reason behind. This shows the importance of qualitative data that would help us to have a closer understanding as to why visitors are behaving in such a way. Understanding the *why* question makes the difference of 99% of the website analysis that is done, yet yielding little insight, and the 1% that provides a picture of the user's mind. Combining the *what* (quantitative) with the *why* (qualitative) can be extremely powerful for providing insights that improve consumer experience.

Many methods are available to collect user-qualitative (*why*) data including,

- Lab usability test (participants are invited to complete tasks, either instructed or non-instructed)
- Site visits, also referred to as follow-me-home (observation of customer's native environment)
- Experimenting
- Unstructured remote conversation (interviewing your real customers when they are visiting your website)
- Surveying

Combining the *what* and the *why* data becomes the cornerstone of any successful actionable analytics program.

8.4 Key Metrics in Web Analytics

Metrics are key concepts with web analytics. They are the numeric measurement of various aspects of user behavior. In this section, we will introduce some basic and key metrics with their implications.

8.4.1 *Visit*

Visit is a single time that a person comes to a website, browses, clicks, and then leaves. In Google Analytics, a visit ends when the user has not viewed any page for 30 minutes. If the user comes back to the website, then it is counted as a new visit.

8.4.2 *Unique Visitors*

When a user starts a new session, analytics counts that as another visit. Unique visitor is a way of capturing how many individuals came to a website and whether they visit multiple times or just once. A visit is unique within the timeframe you have selected; it does not say if you are an old or new visitor to the website.

8.4.3 *Pageviews*

A pageview is a single time that a user visits a page. If the user goes to a page more than once, then it is counted as separate pageviews. For example, in a web journey, if a visitor has landed on five pages and comes back to a repeated page, then pageviews is counted as six. Pageviews of different pages on a website tells you the popularity of those webpages.

8.4.4 *Average Visit Duration*

Average visit duration is the amount of time that a user spends on the website. However, the time a visitor spends on the last page is censored, as it is not observable. The average visit duration is often used as a visitor's engagement indicator with the website.

8.4.5 *Bounce Rate*

The bounce rate of a page or a group of pages is the number of users who exit the website after viewing that page. It counts the number of visitors who end their session with the website after landing on that page. Intuitively, a low bounce rate is good. However, a page with high bounce rate can indicate two very different things. A page with high bounce rate might show the visitors' frustration or unfulfilled goal after visiting that page, thus they are leaving the website. However, it might also be because the visitors have found what they need, thus leaving the website. As we have mentioned in the previous section, metrics alone only tell you *what*, but not the *why*.

8.5 Visitor Analysis

Visitor analysis is the web analytics that allows you to obtain the technical and demographic details about your visitors. In general, it allows you to know

- Where are my visitors from geographically
- Their operating system and browser
- Screen resolution
- Whether or not they use a mobile device to visit the website

8.5.1 *Demographics*

This web analytics tool uses IP address to locate the city where the visitors are from. If graphical information matters to the website, then this report gives you a good proxy of the visitors' geographical distribution.

8.5.2 *Behavior*

A returning visitor is someone who has visited your website before. The behavior report regarding *New vs. Returning visitor* lets you know the percentage of the new visitors to the website in the selected time framework. In this way, you have a good overview if the existing customers are engaged and if new customers are explored and reached.

The *Frequency* and *Recency* report gives you first a breakdown of the number of visitors in a selected time framework who visited the website once, twice, or three times. It also gives you information regarding the number of days since the customer's last visit to the website. The shorter the recency, the stickier the website is to the users.

Finally, the engagement report gives the visit duration and the page depth of the visitors. Again, it is a breakdown of visitors based on their average visiting duration and page depth. Then you may compare across the website pages to find the popular pages vs. unpopular pages.

8.5.3 *Technology*

The technology report allows you to learn the visitors' browser or operation system and the specific version they are using. It also allows you to know the screen resolution of their devices. Then it helps to answer questions such as "if we change the website for a 800 × 600 resolution, what will be the portion of benefited visitors? What will be the portion of the users affected negatively?"

8.6 Traffic Analysis

Traffic analysis allows you to know the source and medium that brings your customer to your website. This is extremely relevant when the marketing team is trying to optimize and measure the effectiveness of their marketing spending across different channels.

A source is a specific place the visitor comes from. Source would be direct, when a visitor types the URL into the address bar of his/her browser or follows a bookmark, and it becomes the tag. It can also be indirect, such as through search engines (organic or paid search) or referral links. In Google Analytics, it can show you the information (keyword) a visitor typed to land on your website. The traffic sources report of Google Analytics provides you with the distribution of visitors across different keywords they used to arrive at your website. Then you may optimize your webpage content so that the page becomes more relevant when accessed by Google Search algorithms.

In terms of referral traffic report, it gives not only the website where the visitor came from but also the page they visited before landing on your webpage. This gives an implication of the activity the visitors are doing before coming to your website and may lead to a significant medium that brings visitors to your webpage.

8.7 Click-path Analysis

The click-path analysis allows the website owner to understand how the users move from one page to another page. However, usually, it is very difficult to characterize a common path for a majority of the users.

Navigation summary report allows us to study the interrelationship of the webpages from the visitors' perspectives. As shown in Figure 8.2, it shows the previous pages before visitors landed on this page and the next page visitors are leaving for. An analysis of the navigation report allows us to understand the relationship between the webpages.

The navigation summary report shows detailed information per webpage. However, it zooms into the links between the webpages and also only a small part of data is reflected on the navigation summary report. The visitor flow report aggregates the data and visualizes it in a meaningful way. As shown in Figure 8.3, the visitor flow report visualizes how the visitors are moving from the first page to the second page to the third page.

Till now our discussion focuses more on analytics between page comparison. In the following section, we will discuss about web scraping, which explains how data can be collected, arranged, and finally analyzed to generate a deep insight regarding the page content.

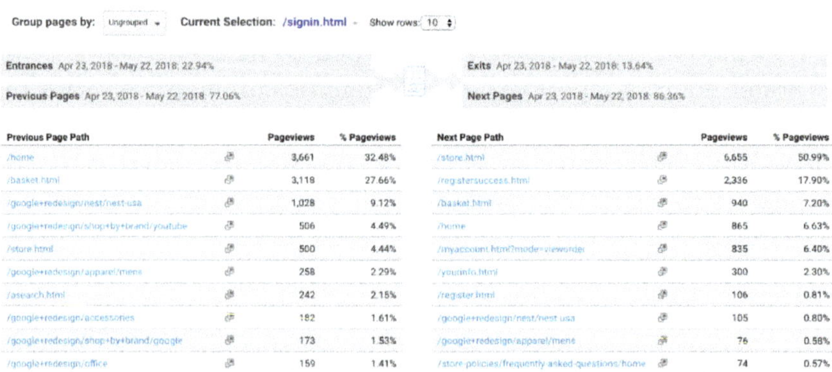

Group pages by: Ungrouped   Current Selection: /signin.html   Show rows: 10

Entrances Apr 23, 2018 - May 22, 2018: 22.94%

Previous Pages Apr 23, 2018 - May 22, 2018: 77.06%

Exits Apr 23, 2018 - May 22, 2018: 13.64%

Next Pages Apr 23, 2018 - May 22, 2018: 86.36%

Previous Page Path	Pageviews	% Pageviews
/home	3,661	32.48%
/basket.html	3,118	27.66%
/google+redesign/nest/nest-usa	1,028	9.12%
/google+redesign/shop+by+brand/youtube	506	4.49%
/store.html	500	4.44%
/google+redesign/apparel/mens	258	2.29%
/asearch.html	242	2.15%
/google+redesign/accessories	182	1.61%
/google+redesign/shop+by+brand/google	173	1.53%
/google+redesign/office	159	1.41%

Next Page Path	Pageviews	% Pageviews
/store.html	6,655	50.99%
/registersuccess.html	2,336	17.90%
/basket.html	940	7.20%
/home	865	6.63%
/myaccount.html?mode=vieworder	835	6.40%
/yourinfo.html	300	2.30%
/register.html	106	0.81%
/google+redesign/nest/nest-usa	105	0.80%
/google+redesign/apparel/mens	76	0.58%
/store-policies/frequently-asked-questions/home	74	0.57%

Figure 8.2: An Example of Google Analytics' Navigation Summary

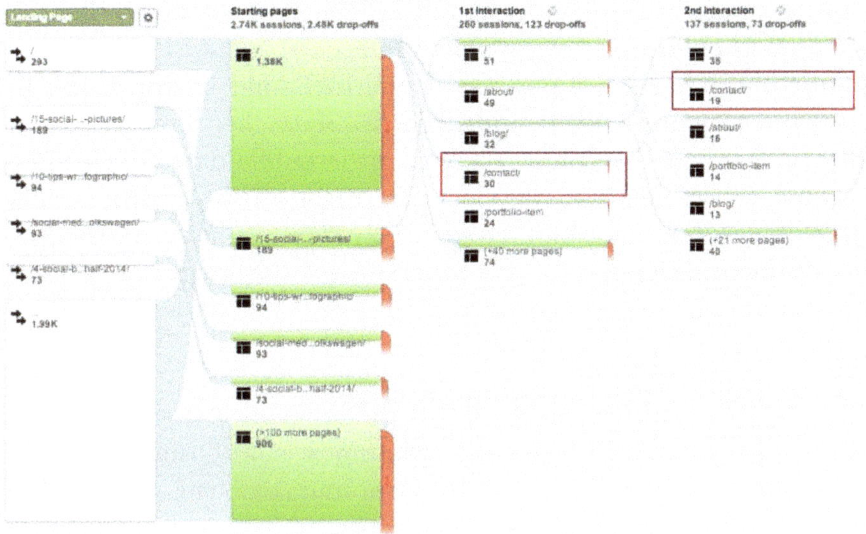

Figure 8.3: An Example of Google Analytics' Visitor Flow Report

8.8 Web Scraping

First of all, what is web scraping? It is quite straightforward by the name of it. Basically, web scraping means the technique to extract information on the webpage and save the information in a way people can easily access and analyze. The process of extracting the data and saving it in the preferable format is effective and automatically done. For example, the web scraper extracts data online and saves it into a CSV file. Then you will be able to access and analyze the data in the way that you want. The web-scraping technique liberates people from copying and pasting the website data manually, which is not feasible for a large scale of webpages. Web scraping simplifies the procedure and speeds it up. This technique is widely used in various industries in today's world with tremendous online information and it has become increasingly important.

Let us take a look at the history of web scraping first. It has gone through quite a long journey until today, dating back to the time when the Internet was born. In the early days, people copied information on a webpage and pasted it to a local file manually.

This approach was extremely slow and ineffective. In 2000, the first web API (Application Programming Interface) and API crawler was invented. API is a set of procedures and tools that make it easier to develop a program with building blocks. Since then, some websites offer APIs so that people can access their public data. Then more software applications were developed to perform web scraping even in those websites that do not offer API. Nowadays, artificial intelligence has also become a powerful tool in scraping and aggregating relevant information online automatically.

8.9 What Is Web Scraping Used for?

Irrespective of whether you are starting a new business, setting strategies for your company or business unit, or being a researcher studying consumer behaviors online, you need to collect and analyze a large amount of data. Most important decisions made today are data driven and evidence based. A vast amount of rich data is available and hidden in the Internet. It is just a question of how do we get it. Web-scraping technique comes into the picture and helps in this angle.

Here, we list down some of the most commonly seen applications in different industries. Because of how fast and effective web scraping can be done, many tasks could not be imagined before this technique was invented.

First, you can monitor every piece of information you are interested in online periodically and you can set the frequency as you like.

For example, monitor the price changes. As an e-commerce website, you might want to be kept updated about your competitors' pricing strategies and react accordingly. But it is too troublesome and even impossible to always go to the competitors' webpage and watch out for any changes as the prices are so dynamic. Or as an airline ticket price comparison website, you need to extract the pricing information of each ticket and keep the pricing information listed on your webpage up to date all the time. This is an impossible task without the web-scraping tool.

Besides competitors' prices, you can fetch other data you want to know about the competitors and make a comparison with your company. The competitive analysis can be easily done with the assistance of web-scraping tools.

Moreover, as a company that cares about the performance of the next product you just launched, you want to understand consumers' feedback about the usage, based on which you can immediately respond if there is an urgent issue or make the improvement in the product generation. You can address consumers' complaints in time, which is very important to keep high satisfaction from the consumers. Or on another occasion, companies want to estimate the sales outcome of the product by monitoring consumers' evaluation. To be able to do so, unlike monitoring the price changes, it requires the extraction of the unstructured data and also the analysis of it. We call it text analysis or sentiment analysis. The technique to be adopted is called natural language processing. We will come back to this later.

The sentiment monitoring can even be used to replace a poll and to predict an election's outcome. It is also used to see how people's sentiment changes (expectations) affect the financial market.

As a researcher, this tool also makes our life much easier. Data used to be a scarcity. Previously, we need to either have resources from companies, purchase the data collected by marketing research firm which is not cheap, or conduct costly surveys to understand the consumers. But now, the web data scraping gives us a new channel to collect the data. By just observing what consumers do and say online, we get to know them at almost zero cost. And the observation can be even more convincing than the answers to the surveys that they are paid to do.

Web scraping can also help with monitoring and understanding the underlying ranking rules of search engines, and based on which, one can adjust their website's content to have a better SEO result.

Second, to aggregate the interested online information. The information that interests you can be scattered and hidden everywhere on the Internet. To search, read all the articles and extract the relevant information, as aggregating the pieces together will take a vast amount of time.

For example, in the information-intensive industry like financial research, people need to be observable of the major events/news that may potentially affect the market and investment decisions. The automatic process helps people to find the relevant ones by keywords and aggregate them into pre-defined categories, which tremendously speeds up the financial analysis work and makes it extremely effective.

Online data aggregation can also be used to verify a person's background and decide on the issuance product to minimize the risk for the company. Background check can also be applied to the human resource (HR) recruitment as a screening process.

Third, real-time analysis and response. This is based on the fact that we can get the real-time data by web scraping. Then, we can also automate the process of analysis and response once the data are available. The real-time analysis and response can be used in the customer relationship management to enhance the satisfaction. It can also be used to generate the real-time prediction of consumers' next step behavior and enhance sales.

In a word, web scraping is a powerful technique that provides solutions to many unsolvable problems people faced in almost all the business domains and beyond 10 years ago. Human errors in data collection are neglected and the efficiency is improved exponentially.

8.10 Web-scraping Procedure and Tools

The easiest way to scrape data online is using the API provided by the websites, although people usually do not consider API as the real web scraping. The process is very simple. Let us use Twitter API to download the tweets with the keyword "trade war" as an example (Figures 8.4 and 8.5).

The first step is authentication. It is required for most of the API usage. You need to create an app in Twitter first and obtain the consumer key, consumer secret, access token, and access token secret. Then you will be able to be authenticated to access the Twitter API

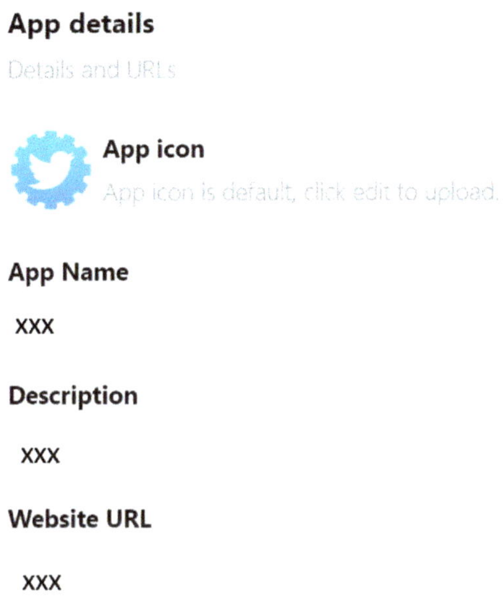

Figure 8.4: Create an Application in Twitter

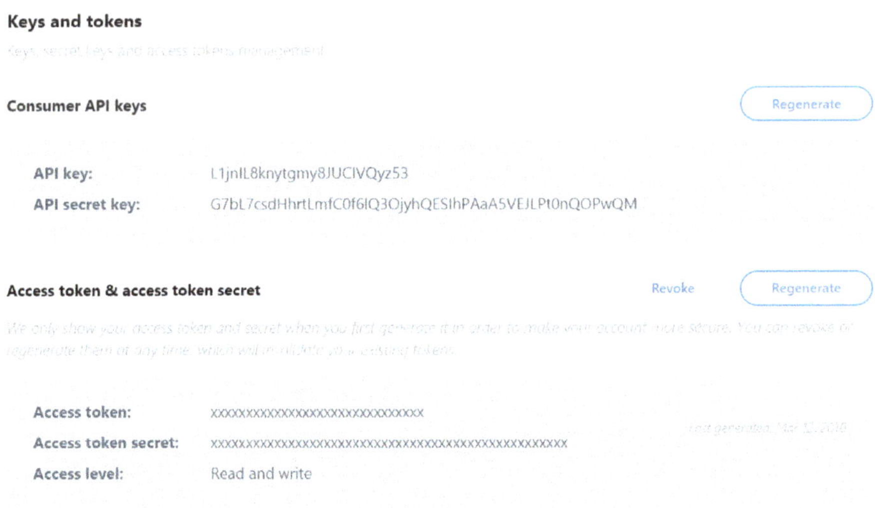

Figure 8.5: Obtain Authentication Keys

using the following R code to complete this step:

```
consumer_key<-‘*****’

consumer_secret<-‘*****’

access_token<-‘*****’

access_secret<-‘*****’

setup_twitter_oauth(consumer_key, consumer_secret, access_token,
access_secret)
```

Then we can obtain the contents of interests:

```
tradewar<-searchTwitter(‘trade war’, n=500, lang="en", resultType
= "recent")
```

Now, let’s talk about the “real” web scraping using XPath language. XPath means XML Path. It is used to navigate via the elements in an XML document. Each item on a webpage can be identified and found by its path. By understanding this and setting a rule to scrape the information following a certain path, one can automatically scrape the data of interest. Here is an example of the XPath language in R code of scraping the book-related information, like name, author, price, etc. on a certain webpage (Figure 8.6).

There are numerous data-scraping software applications in the market, catering to the different needs and requirements. We do not need to do word-by-word coding in the scraping procedure.

For example, Google Chrome has an extension called web scraper, which can do the task beautifully. You can set up a sitemap as shown in Figures 8.7 and 8.8, which tells the logic of how to scrape the data: (1) get all the book items on a webpage, (2) obtain the necessary information like price and name of each book item, and (3) automatically turn pages — repeat steps (1)–(3).

8.11 Text Mining, Sentiment Analysis

After we scrape all the necessary data, the next step is to analyze them. Since the majority of data we scraped from the Internet is text-based, i.e., reviews, articles, news, critiques, etc., it is important to

```
for (item.no in 1 : length(page.list) ) {

    # parse book info
    item.book = page.list[item.no]
    item.child = xml_children(xml_children(item.book)[3])
    item.author = xml_children(xml_children(item.child[2])[2])
    item.pub = xml_children(xml_children(item.child[3])[2])
    item.price = xml_children(xml_children(item.child[5])[2])
    book.name = xml_attr(item.child[1], "title")
    book.author = xml_attr(item.author[1], "title")
    book.pub = xml_attr(item.pub, "title")
    book.sku = xml_attr(item.price[1], "data-price-id")

    # parse book price (from json)
    book.json = paste0("//p.3.cn/prices/mgets?skuIds=J_", book.sku)
    book.json = url_absolute(book.json, page.url)
    book.price =  fromJSON(readLines(book.json, warn="F"))


    # done: we've grabbed all info about the book

    # add book to list
    books$name = c(books$name, book.name)
    books$author = c(books$author, book.author)
    books$pub = c(books$pub, book.pub)
    books$price = c(books$price, book.price$m)
    books$deal = c(books$deal, book.price$p)

    num = num + 1
    cat(" page", page.no, " book", num, ": ", book.name, "\n")
}
```

Figure 8.6: R Code Example Using XPath

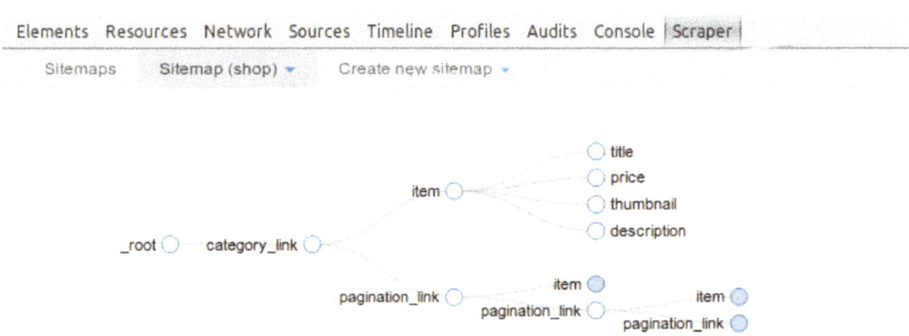

Figure 8.7: Set up Scraping Logic Flow in Web Scraper

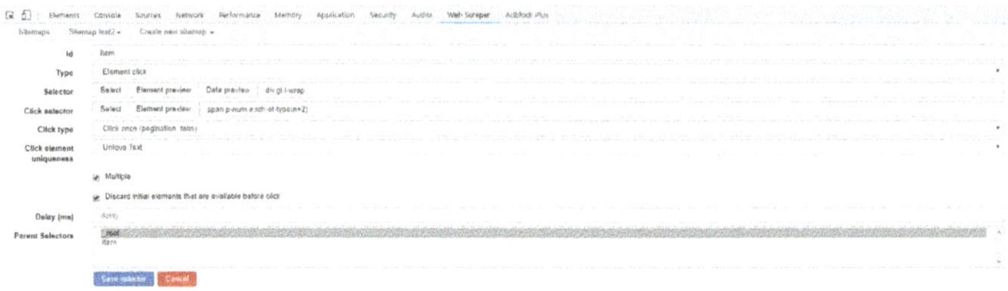

Figure 8.8: Set up Each Selector in Web Scraper

extract quality information, understanding, patterns, and sentiments from the data. Text mining is applying statistical methods and machine learning to solve this problem.

Sentiment analysis, or opinion mining, is one of the most popular applications of text mining. It is widely used in social media listening, to understand people's perception of the newly launched product. Essentially, what sentiment analysis does is to automatically detect the emotion, mood, and feeling expressed by the target sentence or paragraph.

Sentiment analysis adopts the polarity analysis method, meaning assigning scores to the document based on the positive or negative emotions. There are two major techniques to do this.

Ones is using lexicon that contains many words with the assigned values of sentiments. By mapping the target document and the lexicon, one can calculate how many positive and negative words appear in the document.

Another technique is supervised machine learning. The two steps involved are training the model with the training data and evaluating the model with the testing data.

8.12 Cases of Using Web Scraping

I will first use a paper studying the relationship between movie box office revenue and movie online reviews (Chintagunta *et al.*, 2010) to demonstrate how the data scraping is used to collect the data and eventually demonstrate the causal relationship.

The data combine box office information and user ratings of each movie. The online review data were collected from movies available in yahoo.com. The data collected include each user review's user ID, overall rating, ratings for story/acting/direction/visuals, rating data, and the number of people who read the reviews and who found the review helpful. The user ratings were collected for the entire duration of the data.

With the data and empirical modeling, the authors managed to establish that the valence of reviews instead of volume of reviews is the main driver of box office revenue.

Another example is a paper from O'Connor *et al.* (2010) showing the significant correlation between public opinions gathered from polls and the sentiment analysis results based on the posts on Twitter between 2008 and 2009.

The data consist of 1 billion Twitter postings between 2008 and 2009. They used Twitter API and managed to crawl the data. The Twitter messages ranged from 100,000 to 7 million per day. The volume of messages increased a lot during this period of time.

As the Twitter postings are all in texts, the research adopted sentiment analysis to understand the sentiment of each posting. The results showed that the sentiment detector based on Twitter data is very similar to the consumer confidence polls and the presidential job approval polls during this period of time.

Similarly, there is another paper (Bollen *et al.*, 2011) that used similar data-scraping and analysis techniques, which found that the mood of calmness can predict the Dow Jones Industrial Average.

References

Bollen, J., Mao, H., & Zeng, X. (2011). Twitter Mood Predicts the Stock Market, *Journal of Computational Science*, 2(1), 1–8.

Chintagunta, P. K., Gopinath, S., & Venkataraman, S. (2010). The Effects of Online User Reviews on Movie Box Office Performance: Accounting for Sequential Rollout and Aggregation Across Local Markets, *Marketing Science*, 29(5), 944–957.

O'Connor, B., Balasubramanyan, R., Routledge, B. R., & Smith, N. A. (2010). From Tweets to Polls: Linking Text Sentiment to Public Opinion Time Series. In *Fourth International AAAI Conference on Weblogs and Social Media*.

Chapter 9

Business Analytics

9.1 Big Data, Big Gain

In the 1960s, Gordon Moore, co-founder of Intel, formulated what is known as Moore's law, which predicts the speed of computer capacity in processing and storing data. Moore's law states that the number of transistors on a computer chip will double every two years. The fulfilment of Moore's law leads to what is called big data. As said in IBM's "Big Data at the Speed of Business" talk series,[1] we create 2.5 quintillion bytes of data every day, which means 90% of the data we possess have been created in the past two years. In a narrow sense, big data means fast analysis of sheer-sized unstructured data to generate valuable business insights.

In the past, paradigm shifts used to take decades. Now a single decade is like the unit of a paradigm's life. The rise of companies like Google, Amazon, and Facebook shows the power of big data and its ability to change not only the world of business but also the world as a whole. Body-fixed sensors are used to track athletes' performance metrics, such as inter-stroke interval, swim time, and turn time. Dan Siroker, a former Google employee, helped Barack Obama to use data to select the most effective campaign from various plans (Callahan, 2015). In business, the world's largest hotel chain (Airbnb), taxi company (Uber), and retailer (Taobao) are all literally

[1]https://developer.ibm.com/dwblog/2017/open-power-hortonworks-hdp-elastic-big-data/.

data companies. From sports, politics, and business, big data has impacted everything. In fact, a new culture of decision-making is being established. With the presence of data, decision-making should not be solely determined by the HIPPOs — the company's highest paid person's opinion. Instead, the senior executives should be willing to overwrite their intuition when the data do not agree. Such culture is referred to as "muting the HIPPOs".[2]

Business analytics is the practice of iterative, methodical exploration of an organization's data to generate business-related insights. Companies committed to data-driven decision-making use it. Chapter 4 has discussed how analytics and new technologies have been applied in the finance industry. Chapter 8 focuses specifically on web data and web analytics. Another field, where business analytics has played a great role is in the area of marketing. When the marketplace has moved online, consumer's behavior in their shopping journey can be largely stored and analyzed. In this chapter, we will introduce four predominant marketing analytics: consumer choice model, conjoint analysis, segmentation, and customer relationship management. The four topics are logically linked in a sequential order. The part of understanding consumer choice elaborates how the utility function of a consumer is constructed and further explains how the utility function drives a consumer's choice over competing brands. With the understanding of consumer choice, conjoint analysis aids as a tool for marketers to design a product. With the product designed, the section of segmentation discusses how demands can be classified into different clusters. Finally, yet importantly, with purchases, customer relationship management explores how we could value the customer base, to differentiate among customers and to maintain the firm's relationship to reduce the churn rate.

9.2 Understand Consumer Choice

Consumers enjoy the benefits of the product they purchase. Each product can be viewed as a bundle of attributes. For example, a smartphone can be viewed as the bundle of screen size, camera,

[2] https://introit.typepad.com/introit/2012/10/muting-the-hippos.html.

battery life, processing power, storage, touch screen, accessories, platform, etc. Each of these attributes contributes to the consumer's total utility. The key question is which attributes should be included into a product? Which factors are eventually driving consumer choice?

9.2.1 *The Utility Function*

In today's economy, we rarely see any monopolistic market. In each product category, consumers are presented with various competing brands. A consumer's choice between the competing brands at each purchase occasion generates one demand for the winning brand. A product's demand can thus be viewed as the aggregation of individual consumer's brand choice decision. Suppose a consumer i's utility function for a particular product of brand j at time t can be expressed as

$$u_{i,j,t} = \alpha_j + \beta_1 x_{1,i,j,t} + \beta_2 x_{2,i,j} + \beta_3 x_{3,i} + \epsilon_{i,j,t} \tag{1}$$

where in equation (1), $u_{i,j,t}$ is the utility consumer, i will gain from purchasing product of brand j at period t. The x's are the independent variables and stand for different product attributes. When x varies with i, j, t, it means a product attribute that varies from different purchase occasions, for different brands, and at different time, such as price when the product is purchased. When x varies only with i and j, it indicates product attribute that does not vary across time. For example, the taste of the chocolate is relatively stable. While x only varies with i, it means this independent variable only differs across different consumers. Such a variable could be a consumer demographic variable. Though they are not product attributes, they do affect the perceived utility of the product by a consumer. For example, there are cigarette brands targeting female consumers. Such a brand would have larger utility for female consumers vs. male consumers. In equation (1), the β coefficients preceding the independent variables stand for the importance, or the magnitude, of each independent variable in increasing or decreasing the utility of the brand consumed. The intercept α_j is a constant and stands for a brand's intrinsic value to the consumer. Such intrinsic value is often referred to as brand value as well. Last but not least, in equation (1) the $\epsilon_{i,j,t}$ term refers to all the random factors that affect the utility, but not observed (Franses and Paap, 2004).

9.2.2 *Consumer as a Rationale Economic Agent*

Consumers, deemed as a rationale economic agent, will choose the brand that gives them the largest utility. Hence, in their purchase decision, consumers will construct a utility for each brand under their consideration and make a comparison across the brands. Let us say brand j is the chosen brand; the behavior behind the choice can be expressed as

$$u_{i,j} = \max(u_{i,1}, u_{i,2}, \ldots, u_{i,J})$$

meaning utility of a consumer's chosen brand is the maximum in the choice set from brand 1, brand 2,..., brand j.

9.2.3 *The Choice Probability*

A consumer can evaluate each brand under consideration. Hence, the consumer is able to observe the utility attached to each of the brand options, unlike the researcher or analyst who is not able to do so. Since market researchers are not able to observe the utility scores that the consumer attaches to each brand option, researchers could only infer the utility that the consumer is likely to give conditioning on product characteristics and consumer demographics. The consumer knows deterministically the utility score of each brand and thus makes his/her choice. To put it mathematically, the consumer knows his/her choice for brand j with probability equal to 1. For researchers, even with an educated guess, such inference can only be probabilistic. Hence, what a research can conclude is "the probability that the consumer choose brand j is...", namely

$$\Pr[Y_{i,t} = j] = f(u_{i,j,t}) = f(x_{i,j,t}) \tag{2}$$

where $Y_{i,t}$ is consumer i's choice at period t. The probability of a consumer choosing brand j is a function of the utility of that brand. The larger the utility of the brand, the larger the probability that it will be chosen. Hence, utility is a function of both product characteristics and consumer demographics.

We all know that the range of probability is 0 to 1. However, utility can be either negative or positive. For example, price as a variable usually poses negative utility to the consumer, while other performance-enhancing features bring positive utility. Therefore, we will need a function that maps utility into the probability range between (0, 1). This can be achieved with a logistic function (Franses and Paap, 2004) and equation (2) can then be expressed as

$$\Pr[Y_{i,t} = j] = \frac{\exp(u_{i,j,t})}{\exp(u_{i,1,t}) + \exp(u_{i,2,t}) + \cdots + \exp(u_{i,j,t})} \tag{3}$$

The functional form of equation (3) may look complicated at first glance. However, the structure of the function is very clear. It stands for the exponential utility of brand j divided by the sum of exponential utility of all brands in the choice set. Since exponential quantity is positive and the numerator of equation (3) is always smaller than the denominator, the probability for sure falls between 0 and 1. In addition, equation (2) also ensures that the larger the utility of brand j, the higher the probability that brand j will be chosen. As shown in Figure 9.1, if the sum of the exponential utility of all the brands in the choice set is a completed pie, the larger will

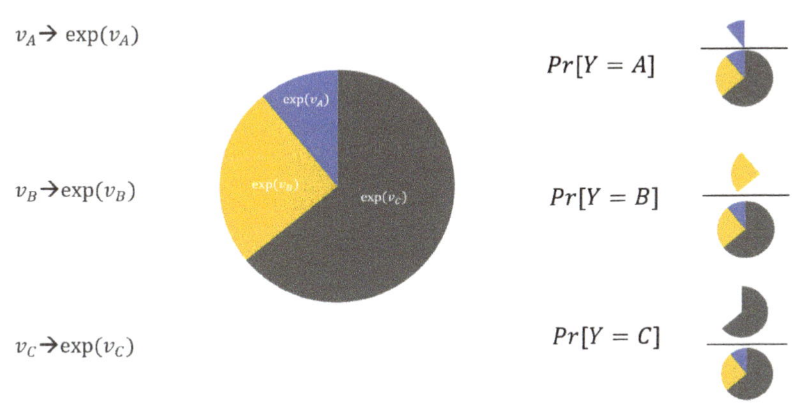

Figure 9.1: The Structure of a Multinomial Logit Model

be the utility of brand j, the bigger the portion it will have in the whole pie and the higher its chance to be chosen.

9.2.4 *Consumer Choice*

The previous sub-section has shown us how to formulate the choice probability for each of the brand considered by a consumer in a purchase decision. In this sub-section, we will elaborate on how choice probability can be linked to consumer choice, the demand, and the market share for the competing brands on the market. The probability of brand j which is chosen can be expressed in equation (3), and we can generate the probability for each brand at the purchase occasion. We can then compare such probability across different brands and safely assume that the consumer will choose brand j if it has the largest utility.

$$Y_{i,j,t} = j \quad \text{if } Pr[Y_{i,t} = j] = \max(u_{i,1}, u_{i,2}, \ldots, u_{i,j})$$

9.2.5 *The Market Share and Demand*

The choice probability of each consumer shows the brand share for a single consumer. For example, a consumer is considering buying a detergent from Tide, Ariel, or Dreft; the choice probability for each of the respective brands is $P_{\text{Tide}} = 0.45$, $P_{\text{Ariel}} = 0.35$, and $P_{\text{Dreft}} = 0.20$. Let us assume this consumer is a representative consumer among 1,000 such consumers for the market under study; we can quite confidently predict the market share among the three brands as $S_{\text{Tide}} = 45\%$, $S_{\text{Ariel}} = 35\%$, and $S_{\text{Dreft}} = 20\%$. Suppose each consumer buys one bottle of detergent in a month, the monthly sales (demand) of each brand would be 450, 350, and 200 bottles.

9.3 Conjoint Analysis for Product Design

As we have learnt in consumer choice, products can be viewed as a bundle of attributes with each attribute contributing to a consumer's expected utility. The question is, when designing a new product, how do we know which attributes are perceived as important factors that will drive purchase, especially when no historical data are available

When choosing a restaurant, how important is ...

	Circle one								
	Not Important								Very Important
Cuisine Type	1	2	3	4	5	6	7	8	(9)
Location	1	2	3	4	5	6	7	(8)	9
Quality of food	1	2	3	4	5	6	7	8	(9)
Price	1	2	3	4	5	6	7	8	(9)

Figure 9.2: Important Factors in Restaurant Selection

Product Option	Cuisine	Distance	Price Range	Preference Rank Value
1	Italian	1km	\$10	1
2	Italian	1km	\$15	3
3	Italian	11 km	\$10	5
4	Italian	11 km	\$15	7
5	Thai	1km	\$10	2
6	Thai	1km	\$15	4
7	Thai	11km	\$10	6
8	Thai	11 km	\$15	8

Figure 9.3: Product Profile Under Conjoint Analysis

for such inferences? One method is to interview consumers to find out the important attribute in influencing their purchase decision. Refer to the survey in Figure 9.2 — when given a list of factors (such as location, cuisine type, food quality, and price) and asked to rate their importance to them while considering a restaurant, most respondents often find it difficult to discriminate between the factors. Thus, most of the responses would fall under the "very important" categories. Conjoint analysis is a method to determine the importance of each factor by forcing respondents to make trade-offs between different product attributes. For example, as shown in Figure 9.3, different product profiles are presented to the respondents. The respondents are required to rank the product profile according to their preference. In this way, the respondents are forced to make a trade-off between different attributes.

Conjoint analysis (CA) is by far one of the most successful and widely used marketing analytics tools. According to Sawtooth Software — a pioneer in CA software development and commercial application and the leading CA consultant — well over 10,000 CA studies are conducted annually. Companies that use conjoint analysis to assist them in their marketing decisions include Bristol-Myers Squibb, Ritz-Carlton, UPS Service, Apple, IBM, and Courtyard by Marriott. A conjoint study can be conducted by following three major steps, which are detailed in the following sub-sections.

9.3.1 *A Study Design*

Conjoint analysis works on the premise that a product can be viewed as a bundle of different attributes. The purpose of the study design (Lilien *et al.*, 2013) is to identify the different attributes to be included in the new or existing product or service. There are also different levels under each attribute. Using different combinations of these attributes, different product profiles are created. The various product profiles are subsequently presented to respondents to rank or rate. We will use the example of sneakers to illustrate how each task is implemented.

9.3.2 *Selection of Attributes*

Decisions need to be made regarding the attributes that will be included for the study. For example, to find out the important factors determining consumers' evaluation of its sneakers, the manufacturer decides that the sneakers' upper, sole, and price are the three most important product attributes. The selection of attributes varies with the purpose of the study. For instance, if the purpose of the study is to understand brand value, then brand perception as an attribute should be included. If the purpose of the study is about repositioning of the product design, then design as an attribute should be included. We will soon explain why it is not realistic to include all the attributes in the study design state.

9.3.3 *Selection of Levels*

Selection of levels refers to the determination of the levels under each attribute. Taking the example of the sneakers, the second attribute,

Attribute	Level Number	Description
Sole	3	Rubber
	2	Polyurethane
	1	Plastic
Upper	3	Leather
	2	Canvas
	1	Nylon
Price	3	$30.00
	2	$60.00
	1	$90.00

Figure 9.4: Study Design for Sneaker Production

upper, could have the three levels — eather, canvas, or nylon. The levels for the first attribute, sole, could be rubber, polyurethane, or plastic. The last attribute, price, could include three different levels — $30, $60, and $90, respectively. With the selection of attribute levels, a table of study design can then be drafted, as shown in Figure 9.4. Every conjoint analysis should have a study design where the attributes and the levels of attributes to be examined are apparent at a glance.

9.3.4 *Creating Product Bundles*

To reiterate, a product can be viewed as a bundle of different attributes. With selected attributes and levels, we can then proceed to make a hypothetical product profile with different combinations of attribute levels, we highlighted the restriction on the number of attributes, as well as the number of levels under each attribute, to be included in a study. The reason for the restriction is that the number of product profile grows exponentially with the increased number of attributes and levels. In our restaurant example, with 3 attributes and 2 levels under each attribute, there are altogether $2 \times 2 \times 2 = 8$ product profiles in total. With our sneakers example, $3 \times 3 \times 3 = 27$ product profiles can be created. Rating or ranking a large number of product profiles can be cognitively challenging for the survey respondents. Thus, having too many product profiles to rate might lead to respondents' evaluation errors and low data quality. A solution to this is to use orthogonal design where only a sufficient number, out of the total number of product profiles, is needed for analysis. The orthogonal design is a balanced design where each level is presented with equal frequency among all the product designs.

9.4 Model and Data

9.4.1 *Designing a Data Collection Method*

With a study design, the firm needs sample representative consumers for their evaluation of the product profiles created. There are three different types of evaluation methods (Lilien *et al.*, 2013) for data collection:

(a) **Pairwise evaluation of product bundles:** Two or more product bundles are presented simultaneously to the respondent who needs to rate on a 100-point scale with a large number denoting strong preference. The drawback with pairwise evaluation is that if we have n product profiles, then the respondent would have to evaluate $n \times (n-1)/2$ pairs. Hence, with 16 product profiles, 120 pairs would need to be evaluated.

(b) **Rank-based evaluation of product bundles:** The respondent ranks the product profile from the most preferred to the least preferred with 1 for the most preferred and 0 for the least preferred.

(c) **Rating-based evaluation of product bundles:** The respondent is presented with all the available product bundles and needs to rate the product bundle on a 100-point scale. The approach assumes that the respondent is able to indicate to what extent they prefer one product bundle over another. With the ease of availability of regression packages across different software, this approach is the most frequently used method.

9.4.2 *Select a Computation Method to Obtain Part-worth Function*

Both pairwise and rank-based evaluations need to employ statistical models such as logit and rank-ordered logit models, which are not within the scope of this chapter. Rating-based evaluation can be applied with a regression model. Hence, we will focus on the details of how we can use regression to obtain the part-worths in the conjoint analysis.

In rating-based evaluation, respondents are asked to rate a score, say between 0 and 100, for a specific product configuration. This score reflects the respondent's preference toward the evaluated product profile. The higher the score, the more preferred the product profile will be. Therefore, we may use this score as the proxy for the product utility for the respondents. The higher the score, the higher the utility of the evaluated product profile will be for the respondents. We use U to denote the utility of the product profile. As the product profile is configured with different attributes, we assume each attribute contributes to the total utility of the product. Hence, the utility of a product profile can be modeled as

$$U(X) = \sum_{i=1}^{m} \sum_{j=1}^{k_i} \alpha_{ij} x_{ij}$$

where $U(X)$ is the overall utility of a product profile, a_{ij} is the part-worth contribution or utility associated with the jth level $(j, j = 1, 2, \ldots k_i)$ of the ith attribute $(i, i = 1, 2, \ldots m)$, $x_{jj} = 1$, if the jth level of the ith attribute is present $= 0$, otherwise, k_i is the number of levels of attribute i, m is the number of attributes.

α_{ij} can be interpreted as how the inclusion of level j of attribute i will lead to α_{ij} increase or decrease in the total utility. Thus, α_{ij} stands for the magnitude of attribute i with level j in changing the total utility, in other words, the worth of this part to the product utility. α_{ij}s is also referred to as part-worths in a conjoint analysis. With any regression package, we may then obtain α_{ij} through estimation.

With part-worths, we may then answer the question about which attribute is the most important to drive consumer choice. The importance of an attribute is defined as I_i, which is the range of the part-worths across all the levels of attribute i. Then the relative importance of this attribute to other attributes can be defined as $W_i = \frac{I_i}{\sum_{i=1}^{m} I_i}$, where $\sum_{i=1}^{m} I_i = 1$.

9.5 Market Simulation

9.5.1 *Segment Customers Based on Their Part-worth*

For each respondent, we can estimate their individual part-worths for each attribute. The part-worth indicates the extent to which the attribute level is valued by the respondents. Hence, the individual part-worths reflect the respondents' preference or usage pattern. For example, in a study of hotel room design, travelers can be classified as business professionals who have smaller part-worths for price and place more value on business amenities as compared to tourists. Based on these part-worths, we can cluster customers into different segments. Often, we see segmentation based on customer demographic information. However, consumer demographics hardly imply any usage preference. In addition, such segmentation can also be easily copied by the competitor. A segmentation based on part-worths is directly linked to user preference. In addition, it is also less tangible and not easily detected by the competitors.

9.5.2 *Design Market Simulation*

With part-worths, we can also calculate the utility of any hypothetical product profile for each respondent. The utility score of competitor product can also be calculated for the same respondents. Assuming that a consumer will choose the product profile that renders him or her the largest utility, we will be able to know the choice of the respective consumers. As stated in Chapter 1, since a product demand is merely the aggregation of its individual consumer choice, we are able to count the demand for each product profile. If the pool of respondents can be viewed as a representative sample of the market, we can obtain the market share for each product profile.

We can also carry out a counterfactual experiment to see how the launch of a new product will change the current market share in the targeted market.

9.6 Segmentation

9.6.1 *The Segmentation, Targeting, and Positioning (S–T–P) Approach*

Consumers have different needs, wants, values, constraints, and incentives to act. In one word, they are heterogeneous. Hence, a diverse range of products or services are required to cater to different consumers. By segmenting the market, firms can better learn consumer needs and wants. They can channel their marketing and product resources into certain segments to serve customers in a more effective and efficient way. Through segmentation, firms operate on a middle ground that do not compel them to rely on a common market that serves all consumers; also, they do not need to incur high cost to customize their marketing effort for each customer. Two dimensions are important when discussing segmentation:

> **Segment:** Refers to a cluster of consumers who have similar needs, wants, and values and are likely to respond to a firm's marketing effort in a similar way.
> **Segmentation:** Refers to the business process to identify, evaluate, and select customer segments that a firm can serve effectively and profitably.

Segmentation is the first step in the STP approach. Segmentation groups customers with similar needs, wants, and values together. Targeting determines the segment a firm chooses to serve, while positioning addresses how well the firm's competitive advantages are compared to other firms which serve the same segment. The STP approach follows five different phases, as listed subsequently. In this chapter, we will focus our discussion on the first two steps of the STP approach.

Segmentation consists of two phases.

Phase 1: Segment the market using basis variables.
Phase 2: Describe the identified segments.

Targeting consists of three phases:

Phase 3: Evaluate the attractiveness of each segment.
Phase 4: Select one or several segments to serve.
Phase 5: Find and reach the customers in the identified segments.

For a segmentation study to be successful, it should lead to a manageable number of target segments with substantial size. It is difficult for companies to manage a few distinctive segments, as each segment needs corresponding marketing effort. It is possible that under each segment, there are microsegments, and it is possible to adjust the company effort to cater to these microsegments. However, the cost to serve these microsegments usually increases quickly. Finally, descriptor variables are often used to describe each segment. For example, variables for online customers might include pages viewed, files downloaded, referrer page, new vs. returning visitors, etc.

A segmentation model requires a dependent variable, and such a variable is called a segmentation basis. A segmentation basis is why customers are classified into different subgroups. As illustrated by panel (a) in Figure 9.5, a selection of a good base of variables (variable 1 and variable 2) is able to divide customers into distinctive subgroups. Likewise, a poor selection of base variables, as shown in panel (b) of Figure 9.5, can hardly organize customers into different subgroups. Hence, the difference between basis and descriptors is that the basis explains how and why customers are different, while the descriptors provide description to customers in each segments.

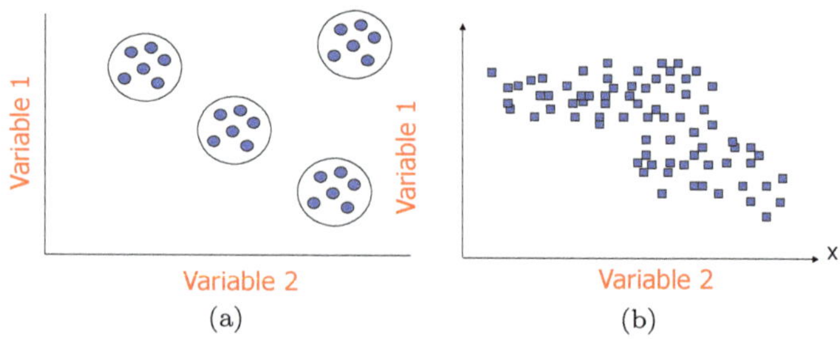

Figure 9.5: Segmentation Based on Base Variables

Demographic information is one type of descriptor. In practice, many descriptors are used as segmentation basis. The problem with segmentation based on descriptors is that they reflect a firm's need to segment customers into different groups, but not customers' preferences or usage patterns. Hence, it is often difficult to develop product or marketing programs that appeal to the customers.

Phase 1: The three basic segmentation methods

A priori segmentation: It is based on industrial knowledge or marketing research; observed characteristics are used to segment customers into different markets, on the assumption that such observed characteristics serve as a good indication of the differences in consumers' needs and wants. For example, many car firms use gender to segment buyers into male and female drivers.

Traditional segmentation methods: These methods develop a single composite based on a set of observable variables. For example, a car company may find customers in its first segment preferring a moderately priced sedan car with good gas mileage, while those in their second segment favor a sporty domestic car that is inexpensive but also with good gas mileage. A further study reveals that the first segment consists of customers with moderate income living in a suburban area, while the second segment consists of young and less affluent people living in the downtown area. The car company decides to denote the first segment as "value seekers" and targets them with traditional sedan cars. The second segment is marked as "thrill seekers" and targeted with sports cars.

Latent class segment: This method is based on a sophisticated statistical model where the difference between the segments is not the mean difference of observable characteristics. For example, in terms of physicians, some of them are patient-focused and some are research-focused. A patient-focused physician might suggest a treatment plan that takes into consideration the patient's medical and non-medical factors. On the other hand, the research-focused physician might implement a treatment plan that reflects the latest research advances. When promoting treatment or drugs to these two different types of physicians, different marketing strategies should be adopted. The firm's challenge, therefore, is to find observables that can explain the physicians' underlying (latent or unobservable) characteristics. In most situations, firms may apply the traditional segmentation approach to gain insights prior to delving into more sophisticated methods.

Phase 2: Segment description

After isolating each segment, the company would need to produce an effective description of them. When describing a segment, the descriptor should indicate the profit potential or the ways in which each segment should be served. Two broad types of variables exist in this sense: one that describes general market characteristics and the other that describes segment insight.

Phase 3: Evaluating segment attractiveness

Firms make their decision to select and serve one or several segments. The attractiveness of each segment can be assessed using the three main dimensions: The first dimension pertains to the size and growth potential of the segment. Though bigger and more accelerated segments are usually preferred by firms, the competition within such segments is also more intensive. Hence, some firms find that it is more profitable to serve the smaller or niche market. The second dimension relates to the structure characteristics of the segment, which include the number, the positioning, and the competitive behavior of competitors; also the saturation of the segment, the protectability (ability to patent, set entry barriers) of the segment, and the environmental risks (political and technological risk). The third dimension pertains to product–market fit. This dimension looks at issues such as determining the presence of coherence between a firm's current image and the new segment it plans to serve, also any presence of synergy between the new segment and its other segments in terms of cost interaction, possibility of cannibalization, and image transfer. Last but not least, the firm cannot neglect the profitability of the segment, the marginal profit, and return on investment.

Phase 4: Select the targeted segments and allocate the resources

After evaluating all the segments, the firm considers the following questions:

(1) How many segments to serve?
(2) Should the firm provide a series of products for one segment?
(3) Should the firm use one offering to serve several segments?

Phase 5: Finding targeted customers

Firms that prefer to serve customers in the right segment need to note that such customers are often reluctant to tag themselves. The only tags associated with them are probably observable information, such as demographic or firmographic (industry, size, location) information. However, what is easy to observe is often a poor predictor for customer value segment. Many segmentation methods clearly identify the segments with different propositions that do not have much of a relationship with demographic or firmographic information. There are two ways that firms can use to identify customers in the right segment.

Customer self-selection: In this case, firms reserve the task and let the consumers select or identify the segment that best matches them. How is this done? The firm provides a range of different products or services that are designed to target different segments and lets the customer choose the product or service that yields him or her the largest utility. Based on the choice of a particular product or service, the customer is segmenting himself or herself.

Scoring/Classification methods: Usually, a set of descriptive items are needed to profile a segment. A major segmentation study can include several hundreds of descriptors. To narrow down the scope, discriminant analysis usually is applied to a small set of variables that can be used to assign the customers to his/her corresponding segment. This method scores the customer for each segment and assigns the customer to the segment in which he or she scores the highest.

9.7 Customer Relationship Management

9.7.1 *From Product-focused to Customer-centric Paradigm*

With increasing competition and the abundant supply of commodities, marketing practice has transformed from a product-focused to a customer-centric paradigm. For instance, instead of asking how profitable a product is, one should ask how profitable a customer is. Instead of asking how one should price the product, ask how the customer should be priced. Rather than ask whether one is doing enough new product development, ask instead if we are developing enough new customers. Ask how we can extend the customer lifetime value instead of extending the product life cycle. Last but not least, consider what our share of customer requirements for individual customer is rather than what our share of the market is. Today, no one will deny the fact that the customer is one of the most important assets of a company. Important questions probably need to be addressed by the marketing manager in each firm — e.g., what the value of the firm's current customer base is, how much to spend on acquiring new customers, which customers to prioritize (i.e., focus the firm's efforts on the most important customers), etc. These questions are all based on customer lifetime value.

9.7.2 *Customer Lifetime Value*

Customer lifetime value (CLV) is the net present value of your **current** customer's current and future contributions to profit. As illustrated by Figure 9.6, we use CLV to evaluate customer relationships in the same way as we use NPV (net present value) to evaluate investments and companies. CLV is the expected NPV of the cash flows from a customer relationship. Mathematically, customer lifetime value can then be defined as

$$CLV = \sum_{t=0}^{T} \frac{m_t \times \prod_{t=1}^{t} r_t}{(1+i)^t} \tag{4}$$

where p_t is the price paid by a consumer at time t, c_t is the direct cost of servicing the customer at time t, m_t is the gross margin from the customer at time t, which is equal to $p_t - c_t$, i is the discount

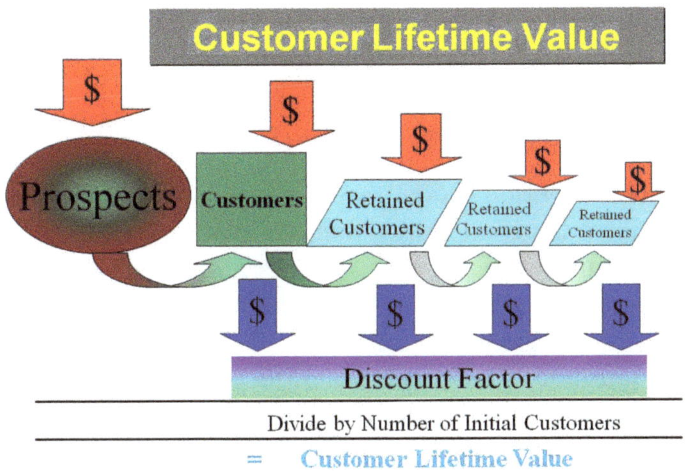

Figure 9.6: Customer Lifetime Value

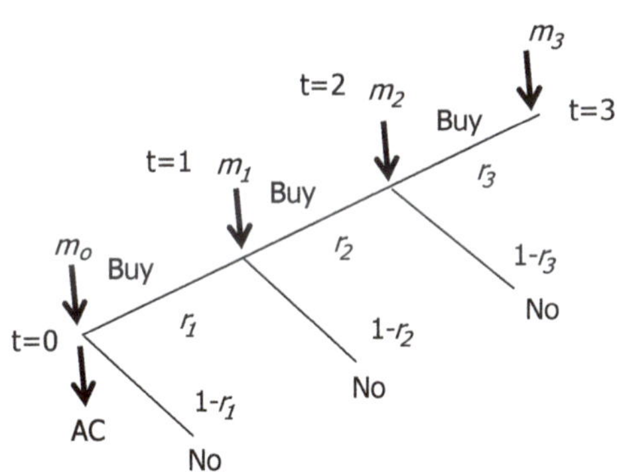

Figure 9.7: Customer Retention Decision

rate or cost of capital for the firm, r_t is the retention rate at time t (the probability that a customer remains a customer), $r_0 = 1$, T is the time horizon for estimating CLV.

As in equation (4), r_t is the likelihood that the customer will stay in period t. Hence, the likelihood that the customer will stay as a customer till period t is then $\prod_{t=1}^{t} r_t = r_1 \cdot r_2 \cdot \ldots \cdot r_t$. The retention decision by a customer can then be presented as in Figure 9.7. In each period, the customer makes a buy or no-buy decision. Should the customer decide not to buy, then the customer is deemed as a lost customer forever. This model is called customer retention model, which is discussed in detail in the following section.

9.8 Customer Retention Model

The customer retention model assumes that customers who are not retained are lost for good. Thus, non-response signals the end of the firm's relationship with the customer. As mentioned in the previous section, this is a simple model to reflect the customer and firm relationship. The customer lifetime value can then be expressed as shown in equation (4). However, when the gross margin m_t as well as the retention rate is the same for each period, then the CLV model can be deduced as

$$CLV = \sum_{t=0}^{\infty} \frac{m \times r^t}{(1+i)^t} = m \times \left(\frac{1+i}{1+i-r}\right)$$

where $\left(\frac{1+i}{1+i-r}\right)$ is called the margin multiplier and CLV is simply the periodic margin times a margin multiplier. We can see from this function that the margin multiplier is larger than 1 by definition. The larger the retention rate r, the smaller the denominator $1+i-r$ and the larger the margin multiplier. Since the discount rate i is very much determined by macrofactors such as inflation rate, the customer lifetime value is mainly driven by the periodic margin m and the retention rate. Thus, to increase an individual customer's profit, we could either increase the periodic profit earned from the customer or enhance his/her retention rate. Cross-selling, which sells multiple products or services to the same customer acquired by the firm, is the marketing practice that increases the profit earned from the customer in each period. In addition, research has also shown that the more types of products a customer buys, the less likely the customer will churn.

9.9 Customer Migration Model

The customer migration model indicates that non-response does not necessarily signal the end of a relationship. The model uses purchase *recency* to predict purchase behavior. In each season, customers have an opportunity to purchase and, based on their past purchase behavior, the purchase propensity of each recency cell is estimated. At the expiration of each period, purchasers "move up" to recency cell number 1 while non-purchasers "age" into the next older recency cell. For example, the famous Jane Doe (Pfeifer and Carraway, 2000) case serves as a good illustration for the customer migration model.

> "ABC company is trying to acquire Jane Doe as its customer. ABC expects to get NC in net contribution on Jane Doe's initial purchase and on each subsequent purchase."
>
> "Purchases are made at most once a period. It is possible that Jane will go silent several periods between purchases. For each period that Jane is considered an 'active' customer, ABC will spend M in efforts to remarket to Jane."
>
> "The probability that Jane will purchase at the end of any period is a function of Jane's recency (If Jane purchased at the end of last period, she is at regency 1 for the current period) and is denoted as p_r, where r is Jane's recency state."

From the above mini case, we can see that this is unlike the retention model where Jane Doe is assumed to be a lost customer if she is inactive for a period. On the contrary, Jane Doe is marked as an inactive customer by the firm. The larger Jane's recency state, the more periods she has not responded. In this model, Jane's state is denoted by her recency and in each period she migrates either way to recency 1 if she made a purchase or the next recency if she did not. Thus, in each period, Jane's retention rate is a function of her recency state. It is intuitive that the larger the recency state, the smaller her probability to stay with the firm. For example, when Jane makes a purchase in the current period, she is in recency 1. For the next period, she has probability p1 to repurchase, or she has $1-p1$ probability of not buying anything. In the latter case, she has $1-p1$ probability move to recency 2 for the next period. Similarly, if Jane Doe's current state is at recency 2, she has p2 probability to make a purchase for the next period, returning her to recency 1. Or, she has $1-p2$ probability of not buying, thus moving to recency 3 state. As we can see, even when a customer has no response, the migration model allows the customer to remain with the company. The migration model is thus a more realistic model than the retention model.

9.10 Markov Chain Model

The customer relationship (Markov Chain) model, or MCM, is a very flexible statistical model that can be used to model both customer retention and customer migration situations. A Markov process is a stochastic process with the following properties:

(a) The number of possible outcomes or states is finite.
(b) The outcome at any stage depends only on the outcome of the previous stage.
(c) The transition probabilities are constant over time.

The Markov chain is a time-homogeneous one where the process is described by a single, time-independent matrix, and the final distribution between the states is called a stationary distribution. Note that there is no assumption on the starting distribution; the chain converges to the stationary distribution regardless of where it begins. This is called the equilibrium distribution of the chain.

The MCM model estimates transition probabilities of a customer in a certain state. Using these transition probabilities, CLV can be estimated as follows:

$$V^t = \sum\nolimits_{t=0}^{T} [(1+i)^{-1}P]^t R$$

where V^T is the vector of expected present value or CLV over the various transition states, P is the transition probability matrix, which is assumed to be constant over time, R is the contribution or margin vector, which is also assumed to be constant over time.

In the MCM model, customer status can be expressed as the probability to switch from one status to another. Such information is summarized in the transition matrix, where no strong assumption is made on the customer. The customer can be a lost, hibernating, or returning one, given the defined transition probability between different states. Hence, the MCM model is the most flexible model to capture customer relationship with the firm.

9.11 Big Data, Big Future

The software, the processors, and the storage to make the business digital are here. The challenge is to find the right people, the right resources, the right routine, and the right culture to make the most of the data benefit. In a survey by the IT industry association CompTIA, almost all the four executives interviewed believed if they could harness big data, their firms will become stronger. However, the rub is almost the same percentage of respondents acknowledged that converting volumes of data into actionable business intelligence has been a challenge.

Some key trends defining the future of business analytics include personalization, integration, and Internet of Things (IoT). Personalization is about serving the right content to the right people at the right time and through the right channel. Marisa Edmund (Callahan, 2015), the Vice President of Edmund Optics, said "If I have one thousand different customer visiting my website, I want them to land on a different home page based on their purchase history, their google search before coming to our websites." While consumers leave their traces everywhere from their viewing history, their product search on Google and whether they opened their email, to their interaction with customer services, these databases remain pretty much separate and do not offer a comprehensive view of the customer. Combining systems in one place to have a holistic view of the customer is the next challenge business analytics need to solve. Technically, Application Programming Interfaces (APIs) and other tools are there to integrate these separate systems. The main issue is to build the team and the process capable of harnessing benefits of integrated data. Fast marketers are building up mobile apps to connect their customers, while slow marketers are still speeding up their website construction. Advanced marketers are using data from IOT. In Brazil, Kimberly-Clark has introduced Huggies with a sensor that alerts parents when the diaper is wet. Amazon's dash button, a button that places orders when it is pressed, precisely collects data of consumer's replenishment behavior.

Big data has transformed numerous industries — retail, entertainment, sports, media, etc. It is coming to every sector of the business whether one likes it or not. Business analytics should follow a dynamic and adaptive process to allow data of new format and from new contexts being analyzed to generate relevant business insights.

References

Callahan, G. (2015). *Big Data-driven Business.* Hoboken: John Wiley & Sons, Inc.

Franses, P. H., & Paap, R. (2004). In *Quantitative Models in Marketing Research* (p. 33). Cambridge: Cambridge University Press.

Lilien, G. L., Rangasways, A., & Bruyn, A. D. (2013). In *Principle of Marketing Engineering* (p. 255). DecisionPro, Inc.

Pfeifer, P. E., & Carraway, R. L. (2000). Modelling Customer Relationships as Markov Chain, *Journal of Interactive Marketing*, 43–55.

Chapter 10

The Power of Digital Inclusion

10.1 Introduction

The world is now becoming increasingly digitalized. In the previous chapters, we have covered various new technologies and business models emerging from the digital economy.

The speed at which the digital transformation has taken place has led to unequal progression among and within societies. Even as many societies progress, there are those who have been left behind. The have-less are often overlooked on the route to the virtues of society's digitization.

As such, a key aspect of digital transformation is the notion of "digital inclusion", the empowerment of individuals through digital participation (Sharma *et al.*, 2016a). Successful initiatives can bring those who are isolated into the digital world and train them to use information and communication technologies (ICTs) to gain social and economic benefits. In today's techno-centric world, technological development has opened more pathways for knowledge sharing and information flow. A networked economy facilitates greater and wider participation of these digital citizens in their countries' socio-economic activities, including many of the traditionally underserved groups. Successful digital inclusion initiatives can enable those who are isolated to gain on the social and economic front.

10.2 From Digital Divide to Digital Inclusion

A digital divide is an economic and social inequality with regard to a gap in access to, use of, or impact of ICTs. Such divide may appear between countries or regions or within a country. Within a country/region, the socio-economic inequalities may appear as technological gaps between individuals, households, businesses, geographic areas, or demographic characteristics. Whereas discussion around the digital divide tends to focus on the access available to individuals, digital inclusion is meant to signal a focus on a practical, policy-driven approach that addresses the needs of communities as a whole.

Looking beyond the idea of a binary divide, Livingstone and Helsper (2007) propose instead a continuum of digital inclusion. Gradations in frequency of Internet use (from non-users and low users through to weekly and daily users) are found to map onto a progression in the take-up of online opportunities among young people (from basic through moderate to broad and then all-round users), thus beginning to explain why differences in Internet use matter, contributing to inclusion and exclusion. Digital inclusion emphasizes on empowering people through ICTs. Many government-initiated digital inclusion activities have been designed to promote the accessibility of ICT and usage for the social and economic development of people with specific needs, including people living in rural areas; persons with disabilities; elderly, women, and children.

10.3 Knowledge Societies

"Mere access" is insufficient to ensure equality of opportunity. The creation of value through knowledge-intensive activities is critical for national growth. It has been shown that knowledge and its diffusion become potent factors of innovation; and this creates a level-playing field of opportunities and hence reduces income and wealth disparities (Piketty, 2014).

> "A knowledge society is a society that is nurtured by its diversity and its capacities. Every society has its own knowledge assets. It is therefore necessary to work towards connecting the forms of

> knowledge that societies already possess and the new forms of development, acquisition and spread of knowledge valued by the knowledge economy model. (UNESCO, 2005, p. 17)"

The creation, development, and governance of knowledge in turn become a pivotal aim of policy-makers in knowledge societies. Prior work highlights key characteristics of knowledge societies: (1) they are necessary and sufficient conditions for growth in the knowledge economy; (2) they have high knowledge-absorptive capacity and complex chains of creation, production, and distribution; and (3) they consist of a sustainable learning community which emphasizes innovation (Sharma *et al.*, 2008).

If knowledge originating from an individual is shared with a mass of people, it could result in the formation of a knowledge society. Access and distribution play a key role in knowledge management.

According to the knowledge assessment methodology (KAM) by World Bank (2008), the four knowledge pillars consist of information infrastructure, economic and institutional governance, education and human capital, and innovation system. Table 10.1 presents the four pillars of the knowledge economy. This methodology has been effective in deriving a set of measures and investigating knowledge policies in several field studies (Sharma *et al.*, 2010).

The four knowledge pillars in the KAM have been effective in informing best practices in the development of knowledge policies (Sharma *et al.*, 2010). A knowledge society generates and shares knowledge that may be used to improve people's well-being to all members of the society.

10.4 Digital Literacy

The term digital divide describes a gap in terms of access to and usage of ICTs. Digital divide had been traditionally considered to be an issue of having or not having access. However, with a worldwide mobile phone penetration of over 95%, it is becoming a relative inequality between those who have more and less digital skills. For example, in the UK, 12.6 million adults lack basic digital skills and 5.9 million have never used the Internet before (BBC, 2016).

Table 10.1: Knowledge Assessment Methodology

Pillar 1	Pillar 2	Pillar 3	Pillar 4
Economic and institutional regime	Education and skills	Information and communication infrastructure	Innovation system
A society's economic and institutional regime must provide incentives for the effective usage of existing and new knowledge.	Members in the society need education and skills that enable them to create, share, and apply knowledge.	A dynamic information infrastructure is essential to facilitate the effective communication, dissemination, and management of information.	The country's innovation system — firms, research centers, universities, think tanks, consultants, and other organizations — must be capable of tapping the growing stock of global knowledge flows, assimilating and adapting it to local needs, and creating new technology.

Source: World Bank (2008).

Table 10.2 summarizes barriers to technology adoption. Urgent action is needed to deal with the digital literacy, or it may risk damaging the societies' productivity and competitiveness. Successful digital inclusion initiatives are those that can bring people who are isolated into the digital space and train them well enough in digital skills to be able to use ICTs to gain social and economic benefits.

Digital literacy is the set of competencies required to fully participate in a knowledge society (Sharma *et al.*, 2016b). It includes knowledge, skills, and behaviors involving the effective use of digital devices such as mobile devices and computers for the purposes of communication, task completion, and collaboration. It also covers usage and comprehension of information in the digital age.

Table 10.2: Barriers to Technology Adoption

Infrastructure	Lack of electricity infrastructure or mobile Internet coverage
User capability	Lack of language literacy or digital literacy
Income and affordability	Low income or consumer purchasing power; cost of devices and telecommunication services
Motivation and incentives	Lack of awareness, relevance, or social acceptance

Source: McKinsey and Company (2014).

On a similar note, Eshet-Alkalai (2004) defines socio-emotionally literate individuals as those who are able to share and evaluate information and knowledge, in order to construct new knowledge. This brings focus to the participation and communication that occurs in the digital world, as well as the opportunities that exist there. Where participation can lead to collective intelligence, this may in turn lead to the development of new knowledge.

Several measures can be taken to promote digital inclusion; (a) by providing free/affordable access to technologies (hardware, software, high-speed Internet connectivity) in their communities; (b) by providing access to a wide range of digital content to their communities; and (c) by providing digital literacy services that assist individuals navigate, understand, evaluate, and create digital content using a range of information and communications technologies; (d) by providing programs and services around key community need areas, such as education, health and wellness, employment and workforce development, and civic engagement.

10.5 Equitable Access Is No Longer Enough

The ubiquity of the Internet poses many challenges and opportunities for individuals and communities. However, these challenges and opportunities have not been evenly distributed. In the past, digital technology has opened new domains of exclusion and privilege for some, leaving some populations isolated from the vast digital realm. Equitable access is no longer enough.

Table 10.3: Stages of Digital Development

Stages	Areas of Focus
Access	Focus on the availability and affordability
Usage	Focus on adoption and usage of digital devices and services
Participation/ Application	Focus on participation in workforce development, education, healthcare, etc., in the digital economy
Value Creation	Focus on human values to enable community involvement, sustainable development, and civic engagement through digitalization

Source: Sharma *et al.* (2016b).

With the emergence of the digital economy, the creation of value through knowledge-intensive activities and emphasis on digital literacy is critical for social and economic growth. Digital inclusion aims to address issues of access, knowledge, skill, and opportunity. It provides a framework for assessing the readiness of societies to provide equitable opportunities in a digital age and raising awareness and sharing information on policies, regulations, and business practices that promote digital inclusion.

10.6 Stages of Digital Development

A networked economy facilitates greater and wider participation of these digital citizens in their countries' socio-economic activities, including many of the traditionally underserved groups.

Digital inclusion has four broad facets: access, usage, participation/application, and value creation. These facets show the ultimate goal of creating digitally inclusive communities.

10.7 Case Studies

10.7.1 *Digital Inclusion and Individuals with Disabilities*

One billion people worldwide live with some form of disability and can only use accessible ICTs. This means that if someone cannot see a typical screen, they need a way to understand what is on it, such as text-to-speech functionality. If a user cannot hear, they need a way to get that information, such as captioning on television or speech-to-text they can read on a mobile phone. Likewise, if someone cannot input a command on an ICT device with their hand, they need alternative ways to do this.

For individuals with disabilities, there are several technologies to support special services to improve the accessibility and quality of life. For people with hearing loss, mobile devices and applications with voice-to-text or text-to-voice functions, visual or vibrating alerts, and devices with the ability to access and control volume can improve their accessibility.

10.7.2 *Digital Inclusion and Women*

In the past, women in many parts of Asia used to face unfavorable conditions with respect to education, healthcare, and employment. As technology advances, the cost of digital service becomes more affordable and mobile devices become more user-friendly. Such connectivity can undoubtedly empower women socially and economically in a number of countries. For example, ICTs provide women with improved access to information on education, job opportunities, and political rights. Digital technologies are also practical tools for women to overcome longstanding economic inequalities. Besides facilitating greater knowledge flows to where such knowledge is most needed, technological platforms can help women purchase the latest technological products/tools through affordable e-commerce channels, obtain cost-effective health services and education, get accessible financial services, and even start their own business online to increase income. According to the report "The Social Wisdom of Wired Women Around the World" by MSLGROUP (2013), more than half of the women surveyed in the US, UK, Brazil, and in urban areas of China said technology had "made their lives more manageable", and nearly 80% of them go to the Internet first when seeking information. Some of these traditionally underserved groups are now able to harness the power of digital platforms and play an important role in the economic and social activities.

10.7.3 *Alibaba's Rural Taobao Strategy*

According to the World Bank data, 43% of the total Chinese population are still living in rural areas. Providing access to e-commerce and digital financial services to millions of the country's poorer residents and micro, small, and midsize businesses (MSMEs) can help to raise the standard of living in China's countryside. It could also be translated into a lucrative opportunity for the e-commerce giants like Alibaba.

The Alibaba Group was listed in the New York Stock Exchange on September 19, 2014. After its initial public offering (IPO), it announced the three major strategies, "e-commerce in rural areas, globalization and big data". Thereafter, the rural strategy of Alibaba Group began to officially come into being. Taobao is Alibaba's e-commerce arm, China's largest e-commerce website with a consumer focus. Alibaba's Rural Taobao Strategy comprises "dual cores", namely, Rural Taobao and Taobao Villages.

To better serve the rural residents and to support agricultural innovation and economic development in rural areas, the Alibaba Group released the "1,000 counties and 10,000 villages" program (namely the rural Taobao mode) in October 2014. This program brings forward RMB 10 billion (about US$1.48 billion) of investment on logistics, hardware, and training in the next three to five years to establish 1,000 county-level service centers and 100,000 service stations in rural areas. The program would develop rural Taobao by building county-level operation centers in county seats and village-level service stations in villages, as part of a rural e-commerce service system. It would build information and logistic channels to "bring consumption goods to the countryside" and explore the ways of selling rural products online.

Rural Taobao: Since three quarters of the Chinese rural population does not have access to the Internet, Alibaba decided to take the services to the customer's doorsteps. Many Taobao Rural Service Centers are located at the local convenience store, where computers and wall-mounted flat-screen monitors are provided, with well-trained villagers serving as representatives. In the service centers, villagers can check out the latest Taobao online deals on mobile phones, toothpastes, pesticide dispensers, and many more. They can get accustomed to making purchases and paying bills online, as well as picking up items they purchased from Taobao. By September 2016, more than 16,000 Taobao Rural Service Centers were already in place. To ensure timely delivery of purchases, Alibaba is opening warehouses and working with delivery companies and local officials.

Historically, with low incomes, dispersed populations, and poor logistic infrastructure, the rural areas have not attracted many brick-and-mortar stores. Thus, shopping in rural China had been characterized by limited choice, inflated prices, and poor quality in the past. With the help of Taobao Service Centers, it takes little time for the rural consumers to realize that the price is cheaper, the choice is better, and it is far more convenient to shop online, noting that they no longer need to make a half-day trip to the dealers. In 2015, the best-selling items sold through Taobao Rural Service Centers were home appliances, mobile top-up, and women's apparel. In this process, Alibaba is nurturing the culture and habit of shopping online among the rural population.

In addition, the centers have provided other innovative services. By working with local branches of Railcom, China Unicom, Telecom, and other operators, the centers provide recharging services and Internet access; through collaboration with Alitrip, they provide services such as train ticket, air ticket, and hotel booking to villagers; through cooperation with Alipay, it grants credit to rural Taobao partners and provides services such as living expenses payment and cash withdrawing in a small sum to villagers. It will soon rely on the platform of Alijk in the future and provide services such as registration, medicine pick-up, and remote diagnosis. On the November 11 shopping festival in 2015, service centers of rural Taobao in the whole country accomplished the turnover of RMB 293 million (about US$43.4 million) in a single day, more than RMB 30,000 (about US$4,400) in each village on average, which demonstrated the huge consumption potential of rural areas in e-commerce.

Taobao Villages: Besides bringing consumption goods to the countryside, Taobao Service Centers also serve another important function — selling the agricultural products to cities. Since 2009, clusters of rural online entrepreneurs who have opened shops on Taobao began to emerge in China. These clusters are often referred to as "Taobao Villages". The number of Taobao Villages has been on the rise since then. According to AliResearch, a "Taobao Village" is defined as a village with a large number of online merchants who adopt Taobao as the key trading platform, relying on the Taobao e-commerce ecosystem and forming a scalable cluster. The criteria of Taobao villages include the following three principles:

1. The trading places are mainly rural areas.
2. The annual turnover in e-commerce reaches RMB 10 million (about US$1.48 million) or more.
3. The number of active online merchants in the village reaches 100 or more, and/or active online shops account for 10% or more of the local households.

In May 2015, Taobao Villages launched the "2.0 mode". Its partners changed from owners of traditional village convenience stores to specialized "rural Taobao partners". The rural Taobao partner program targets the local villagers who are open-minded and familiar with the Internet and online shopping, especially young migrant workers who are returning to their hometown from the cities. A total of 1,311 Taobao Villages widely distributed in 18 provinces were founded in China at the end of August 2016. Apparel, furniture, and shoes are the best-selling product categories sold by the Taobao Village merchants.

With the help of more and more rural Taobao partners, many traditional farming communities have gradually transformed into e-commence clusters with many new job opportunities, including both Internet-based jobs and logistics-related positions. The drain of a young workforce in the countryside used to pose a great threat to the economic and social development of rural areas. By attracting more and more young migrant workers to come back to Taobao Villages from the cities, Alibaba's rural strategy not only promotes business startup opportunities, employment, and economic development but may also mitigate many of the issues caused by rapid urbanization.

The connection to urban consumers through Taobao Villages is also changing the farmers' attitudes toward the goods they produce and their agricultural practices. They earn more by removing the middlemen and cutting short the supply chain, going straight to the consumers. Their consumers are mostly urban residents who are concerned about food quality and safety. Thus, villagers are focusing more and more on producing quality goods rather than cutting cost and compromising on quality.

To support the Rural Taobao Service Centers and Taobao Villages, Alibaba has helped rural areas build the infrastructure of e-business, including trade, logistics, digital payment and financing, cloud computing, data analytics, and so on. Other business entities and entrepreneurs outside of the Alibaba ecosystem can also make use of the infrastructure in the future and contribute to the rural economy. Alibaba is working with third-party logistic service providers to set up the logistic channels in rural areas through subsidy and other means; Cainiao network built the "distribution network for big household appliances", covering 95% districts and counties in the country and 500,000 villages; Mantianxing program had incorporated 51 counties in the country by December 2015 and traced the sources of high-quality agricultural products; Ant Financial had connected more than 2,300 rural financial institutions, served more than two million rural e-businesses, and provided business loans to 180,000 small and microcorporations in rural areas, lending RMB 30 billion (about US$4.43 billion) in total.

Source: Ding *et al.* (2017).

10.7.4 *Digital Inclusion in Singapore*

The Infocomm Development Authority of Singapore's (IDA) Digital Inclusion program is aimed at building a digitally inclusive society that brings Infocomm to people from all walks of life. Committed to growing Singapore into a dynamic global Infocomm hub, the Digital Inclusion program consists of four key areas, namely the

Silver Infocomm Initiative, *NEU PC Plus Programme*, *Home Access Programme*, and *Enable IT Programme* which target four key groups: seniors, needy students, low-income households, and people with disabilities.

Silver Infocomm Initiative: The Silver Infocomm Initiative (SII) aims to bridge the digital divide among seniors aged 50 and above through addressing their differences in educational background, language, and Infocomm competencies. The initiative comprises the following programs:

- Silver IT Fest — Formerly known as Silver Infocomm Day, SID is the flagship annual event that began in 2007. This event aims to enlighten seniors on the latest in Infocomm through seminars, IT workshops, and exhibitions. The total participation attained through the Silver IT Fest series since its inception stands at more than 42,000.
- Silver IT Care — Started in late 2013, the program taps on tech-savvy seniors to help other seniors. The outcome is the setup of the Infocomm Helpdesk manned by senior volunteers who are able to guide their peers with any software or hardware issues. As of December 2015, about 740 cases inclusive of technical queries and computer repairs were successfully dealt with.
- Silver Infocomm Wellness Ambassadors (SIWA) — Now in its fourth year, the program recognizes seniors who have not only embraced Infocomm but also inspired their peers to pick up IT skills. Till date there are 102 recipients of the award.
- Silver Infocomm Hotspots (SIH) — To encourage Infocomm usage among interested seniors with no access to computers, these Infocomm access points were set up at places frequented by seniors. Currently, there are about 90 such SIHs.
- Silver Infocomm Junctions (SIJ) — They are senior-friendly Infocomm-learning hubs with affordable and customized training courses that cover a range of topics from basic Infocomm skills such as using a computer to more advanced topics, such as online storage and sharing (Cloud Computing).
- Intergenerational IT Bootcamp — The program seeks to foster bonds between seniors and their grandchildren or students, while

allowing them to learn Infocomm in a fun way. The program was developed in 2010 and has seen over 2,800 seniors pick up IT skills. As of December 2015, close to 106,000 training places had been achieved through courses conducted at Silver Infocomm Junctions, Silver IT Fest, and Intergenerational IT Bootcamps.

NEU PC Plus Programme — This program aims to build an all-inclusive digital society where students and people with disabilities from low-income households can have equal access to Infocomm by defraying the cost of computer ownership and Internet access. The two schemes under the NEU PC Plus Programme are as follows:

- **PC-Bundle Scheme** — This scheme is applicable to students whose gross monthly household income does not exceed S$3,000 or per capita income does not exceed S$800. The scheme offers a new computer (desktop or laptop) together with a three-year free broadband access and software at an affordable price through a two-tier subsidy for Singapore Citizens and Permanent Residents. Depending on their monthly household income or per capita monthly income (PCI), the funding scheme could be up to 50% or 75%.
- **iNSPIRE Fund Scheme** — This scheme, introduced in November 2007, helps NEU PC Plus applicants who cannot afford the co-payment to earn their PC by doing some form of community service.
- **Home Access Programme** — The Home Access Programme, launched in November 2014, aims to make Internet connectivity more affordable to the low-income households with access to the Internet. This includes the provision of basic mobile computing devices and Internet connectivity in the homes of low-income households. As of March 2016, more than 4,400 had been approved to benefit from the program.

Enable Persons with Disabilities through Infocomm and Assistive Technology (Enable IT) Programme: The program aims to enable people with disabilities through the adoption of Infocomm and assistive technology (IT/AT) to enhance their abilities in the activities of daily living, education, and employment, as well as to help them interact with others. In addition, IDA is collaborating with relevant partners from the disability sector to promote and raise awareness on the benefits of IT/AT adoption to people with disabilities, caregivers, and disability professionals. Guided by the Smart Nation vision, the program is implemented through close collaboration between IDA, disability sector organizations, stakeholders (e.g., policy-makers from government agencies, disability voluntary welfare organizations, and help agencies), and private sector organizations. Through the program's initiatives, people with disabilities will be able to make informed decisions about appropriate technological solutions to adopt. People with disabilities and professionals from the disability sector will also be equipped with relevant IT/AT skills via hands-on experience and training programs. By equipping people with disabilities with technological skills, their employability will be enhanced. In turn, they will be enabled to pursue their dreams, attain a sense of self-worth and financial independence, as well as contribute to Singapore's Smart Nation vision.

Source: Infocomm Development Authority of Singapore (IDA) (2017).

10.8 Conclusion

As of today, the key challenge remains in closing the digital divide and promoting digital inclusion. Countries around the world are making continuing progress on digital inclusion regarding the development of national technology strategies and advancements toward mobile platform interoperability. Several priority action areas have been identified for public sector, private sector, and civil society stakeholders aiming to augment digital inclusion: (1) establishing and achieving specific, measurable digital inclusion targets; (2) promoting more comprehensive data collection and analysis regarding technology access and usage, particularly among traditionally underserved groups, such as the elderly; (3) advancing regulatory efforts designed to facilitate business models and technological capability with a social impact; and (4) enhancing digital literacy among customers in order to promote sustainable digital inclusion. With more societies committed to the above efforts, the gaps arising from the uneven distribution of resources throughout the world can be addressed so that the rich opportunities afforded by the expanding digital landscape are made available to all.

References

BBC (2016). UK Facing 'Digital Skills Crisis' Warn MPs, *BBC News*. Retrieved from https://www.bbc.com/news/business-36510266.

Ding, D., Guan, C., Lee, D. K. C., Tan, L. C. (2017). From Ant Financial to Alibaba's Rural Taobao Strategy — How Fintech Is Transforming Social Inclusion, in Lee, D. & Deng, R. (Eds.), *Handbook of Blockchain, Digital Finance, and Inclusion*, vol. 1, Elsevier Academic Press.

GSMA (2014). *Digital Inclusion Report*. Retrieved from https://www.gsma.com /mobilefordevelopment/

Eshet-Alkalai, Y. (2004). Digital Literacy: A Conceptual Framework for Survival Skills in the Digital Era, *Journal of Educational Multimedia and Hypermedia*, 13(1), 3–13.

Infocomm Development Authority of Singapore's (IDA) (2017). Digital Inclusion. Retrieved from https://www.imda.gov.sg/community/consumer-education/ digital-inclusion

Livingstone, S. and Helsper, E. (2007). Gradations in Digital Inclusion: Children, Young People and the Digital Divide, *New Media & Society*, 9(4). 671–696.

McKinsey and Company (2014). *Offline and Falling Behind: Barriers to Internet Adoption*. Retrieved from https://www.mckinsey.com/industries/technolog y-media-and-telecommunications/our-insights/offline-and-falling-behind-b arriers-to-internet-adoption

MSLGROUP (2013). *Women Online: The Social Wisdom of Wired Women around the World*. Retrieved from http://mslgroup.com/news/2013/20131 112-wired-women.aspx.

Piketty, T. (2014). *Capital in the Twenty-First Century*, President and Fellows of Harvard College.

Sharma, R., Chandrasekar, G. and Vaitheeswaran, B. (2010). A Knowledge Framework for Development — Empirical Investigation of 30 Societies, *International Journal of Knowledge Management*, 6(4), 1–23.

Sharma, R., Fantin, A.-R., Prabhu, N., Guan, C. and Dattakumar, A. (2016a). Digital Literacy and Knowledge Societies: A Grounded Theory Investigation of Sustainable Development, *Telecommunications Policy*, 40(7), 628–643.

Sharma, R. S., Lin, M., Guan, C. and Dattakumar, A. (2016b). A Maturity Model for Digital Literacies and Sustainable Development, In Khosrow-Pour, M. (Ed.), *Encyclopedia of Information Science and Technology*, Fourth Edition, IGI Global.

Sharma, R. S., Ng, E. W. J., Dharmawirya, M., and Lee, C. K. (2008). Beyond the Digital Divide: A Conceptual Framework for Analyzing Knowledge Societies, *Journal of Knowledge Management*, 12, 151–164.

UNESCO (2005). Towards Knowledge Societies: UNESCO World Report. Retrieved from https://unesdoc.unesco.org/ark:/48223/pf0000141843

World Bank (2008). Knowledge for Development, World Bank Institute, Knowledge for Development (K4D) Program. Retrieved from http://siteresources.worldbank.org/KFDLP/Resources/461197-1199907090464/k4d_bookletjune2008.pdf

Index

Singapore University of Social Sciences - World Scientific Future Economy Series

(Continued from page ii)

Blockchain Foundational Issues and Beyond
Lo Swee Won and David Lee Kuo Chuen

Inclusive Disruption: Digital Capitalism, Deep Technology and Trade Disputes
David Lee Kuo Chuen and Linda Low

Lightning Source UK Ltd.
Milton Keynes UK
UKHW021128240520
363702UK00001B/17

9 789811 203923